Reviews of Environmental Contamination and Toxicology

VOLUME 254

More information about this series at http://www.springer.com/series/398

Reviews of Environmental Contamination and Toxicology Volume 254

Editor
Pim de Voogt

Volume 254

Springer

ISSN 0179-5953 ISSN 2197-6554 (electronic)
Reviews of Environmental Contamination and Toxicology
ISBN 978-3-030-68529-4 ISBN 978-3-030-68530-0 (eBook)
https://doi.org/10.1007/978-3-030-68530-0

This Springer imprint is published by the registered company Springer Nature Switzerland AG
The registered company address is: Gewerbestrasse 11, 6330 Cham, Switzerland

Foreword

International concern in scientific, industrial, and governmental communities over traces of xenobiotics in foods and in both abiotic and biotic environments has justified the present triumvirate of specialized publications in this field: comprehensive reviews, rapidly published research papers and progress reports, and archival documentations These three international publications are integrated and scheduled to provide the coherency essential for nonduplicative and current progress in a field as dynamic and complex as environmental contamination and toxicology. This series is reserved exclusively for the diversified literature on "toxic" chemicals in our food, our feeds, our homes, recreational and working surroundings, our domestic animals, our wildlife, and ourselves. Tremendous efforts worldwide have been mobilized to evaluate the nature, presence, magnitude, fate, and toxicology of the chemicals loosed upon the Earth. Among the sequelae of this broad new emphasis is an undeniable need for an articulated set of authoritative publications, where one can find the latest important world literature produced by these emerging areas of science together with documentation of pertinent ancillary legislation.

Research directors and legislative or administrative advisers do not have the time to scan the escalating number of technical publications that may contain articles important to current responsibility. Rather, these individuals need the background provided by detailed reviews and the assurance that the latest information is made available to them, all with minimal literature searching. Similarly, the scientist assigned or attracted to a new problem is required to glean all literature pertinent to the task, to publish new developments or important new experimental details quickly, to inform others of findings that might alter their own efforts, and eventually to publish all his/her supporting data and conclusions for archival purposes.

In the fields of environmental contamination and toxicology, the sum of these concerns and responsibilities is decisively addressed by the uniform, encompassing, and timely publication format of the Springer triumvirate:

Reviews of Environmental Contamination and Toxicology [Vol. 1 through 97 (1962–1986) as Residue Reviews] for detailed review articles concerned with any aspects of chemical contaminants, including pesticides, in the total environment with toxicological considerations and consequences.

Bulletin of Environmental Contamination and Toxicology (Vol. 1 in 1966) for rapid publication of short reports of significant advances and discoveries in the fields of air, soil, water, and food contamination and pollution as well as methodology and other disciplines concerned with the introduction, presence, and effects of toxicants in the total environment.

Archives of Environmental Contamination and Toxicology (Vol. 1 in 1973) for important complete articles emphasizing and describing original experimental or theoretical research work pertaining to the scientific aspects of chemical contaminants in the environment.

The individual editors of these three publications comprise the joint Coordinating Board of Editors with referral within the board of manuscripts submitted to one publication but deemed by major emphasis or length more suitable for one of the others.

Coordinating Board of Editors

Preface

The role of *Reviews* is to publish detailed scientific review articles on all aspects of environmental contamination and associated (eco)toxicological consequences. Such articles facilitate the often complex task of accessing and interpreting cogent scientific data within the confines of one or more closely related research fields.

In the 50+ years since *Reviews of Environmental Contamination and Toxicology* (formerly *Residue Reviews)* was first published, the number, scope, and complexity of environmental pollution incidents have grown unabated. During this entire period, the emphasis has been on publishing articles that address the presence and toxicity of environmental contaminants. New research is published each year on a myriad of environmental pollution issues facing people worldwide. This fact, and the routine discovery and reporting of emerging contaminants and new environmental contamination cases, creates an increasingly important function for *Reviews*. The staggering volume of scientific literature demands remedy by which data can be synthesized and made available to readers in an abridged form. *Reviews* addresses this need and provides detailed reviews worldwide to key scientists and science or policy administrators, whether employed by government, universities, nongovernmental organizations, or the private sector.

There is a panoply of environmental issues and concerns on which many scientists have focused their research in past years. The scope of this list is quite broad, encompassing environmental events globally that affect marine and terrestrial ecosystems; biotic and abiotic environments; impacts on plants, humans, and wildlife; and pollutants, both chemical and radioactive; as well as the ravages of environmental disease in virtually all environmental media (soil, water, air). New or enhanced safety and environmental concerns have emerged in the last decade to be added to incidents covered by the media, studied by scientists, and addressed by governmental and private institutions. Among these are events so striking that they are creating a paradigm shift. Two in particular are at the center of ever increasing media as well as scientific attention: bioterrorism and global warming. Unfortunately, these very worrisome issues are now superimposed on the already extensive list of ongoing environmental challenges.

The ultimate role of publishing scientific environmental research is to enhance understanding of the environment in ways that allow the public to be better informed or, in other words, to enable the public to have access to sufficient information. Because the public gets most of its information on science and technology from internet, TV news, and reports, the role for scientists as interpreters and brokers of scientific information to the public will grow rather than diminish. Environmentalism is an important global political force, resulting in the emergence of multinational consortia to control pollution and the evolution of the environmental ethic. Will the new politics of the twenty-first century involve a consortium of technologists and environmentalists, or a progressive confrontation? These matters are of genuine concern to governmental agencies and legislative bodies around the world.

For those who make the decisions about how our planet is managed, there is an ongoing need for continual surveillance and intelligent controls to avoid endangering the environment, public health, and wildlife. Ensuring safety-in-use of the many chemicals involved in our highly industrialized culture is a dynamic challenge, because the old, established materials are continually being displaced by newly developed molecules more acceptable to federal and state regulatory agencies, public health officials, and environmentalists. New legislation that will deal in an appropriate manner with this challenge is currently in the making or has been implemented recently, such as the REACH legislation in Europe. These regulations demand scientifically sound and documented dossiers on new chemicals.

Reviews publishes synoptic articles designed to treat the presence, fate, and, if possible, the safety of xenobiotics in any segment of the environment. These reviews can be either general or specific, but properly lie in the domains of analytical chemistry and its methodology, biochemistry, human and animal medicine, legislation, pharmacology, physiology, (eco)toxicology, and regulation. Certain affairs in food technology concerned specifically with pesticide and other food-additive problems may also be appropriate.

Because manuscripts are published in the order in which they are received in final form, it may seem that some important aspects have been neglected at times. However, these apparent omissions are recognized, and pertinent manuscripts are likely in preparation or planned. The field is so very large and the interests in it are so varied that the editor and the editorial board earnestly solicit authors and suggestions of underrepresented topics to make this international book series yet more useful and worthwhile.

Justification for the preparation of any review for this book series is that it deals with some aspect of the many real problems arising from the presence of anthropogenic chemicals in our surroundings. Thus, manuscripts may encompass case studies from any country. Additionally, chemical contamination in any manner of air, water, soil, or plant or animal life is within these objectives and their scope.

Manuscripts are often contributed by invitation. However, nominations for new topics or topics in areas that are rapidly advancing are welcome. Preliminary communication with the Editor-in-Chief is recommended before volunteered review manuscripts are submitted. *Reviews* is registered in WebofScience™.

Inclusion in the Science Citation Index serves to encourage scientists in academia to contribute to the series. The impact factor in recent years has increased from 2.5 in 2009 to 7.0 in 2017. The Editor-in-Chief and the Editorial Board strive for a further increase of the journal impact factor by actively inviting authors to submit manuscripts.

Amsterdam, The Netherlands
February 2020

Pim de Voogt

Contents

List of Contributors

Ângela Almeida Biology Department and CESAM, University of Aveiro, Aveiro, Portugal

Tamzin A. Blewett University of Alberta, Department of Biological Sciences, Edmonton, AB, Canada

Williams Kweku Darkwah Key Laboratory of Integrated Regulation and Resource Development on Shallow Lakes, Ministry of Education, College of Environment, Hohai University, Nanjing, People's Republic of China

Valdemar I. Esteves Chemistry Department and CESAM, University of Aveiro, Aveiro, Portugal

Ignacio Fernández-Olmo Dpto. de Ingenierías Química y Biomolecular, Universidad de Cantabria, Santander, Cantabria, Spain

Erik J. Folkerts University of Alberta, Department of Biological Sciences, Edmonton, AB, Canada

Rosa Freitas Biology Department and CESAM, University of Aveiro, Aveiro, Portugal

Greg G. Goss University of Alberta, Department of Biological Sciences, Edmonton, AB, Canada

Laura Guimarães CIIMAR – Interdisciplinary Centre of Marine and Environmental Research, University of Porto, Matosinhos, Portugal

Huang Jingyu Key Laboratory of Integrated Regulation and Resource Development on Shallow Lakes, Ministry of Education, College of Environment, Hohai University, Nanjing, People's Republic of China

Li Ke Department of Civil Engineering, Jilin Jianzhu University, Jilin, People's Republic of China

Dongwei Liu Department of Nephrology, The First Affiliated Hospital of Zhengzhou University, Zhengzhou, P. R. China

Research Institute of Nephrology, Zhengzhou University, Zhengzhou, P. R. China

Research Center for Kidney Disease, Zhengzhou, Henan Province, P. R. China

Key Laboratory of Precision Diagnosis and Treatment for Chronic Kidney Disease in Henan Province, Zhengzhou, P. R. China

Core Unit of National Clinical Medical Research Center of Kidney Disease, Zhengzhou, P. R. China

Zhangsuo Liu Department of Nephrology, The First Affiliated Hospital of Zhengzhou University, Zhengzhou, P. R. China

Research Institute of Nephrology, Zhengzhou University, Zhengzhou, P. R. China

Research Center for Kidney Disease, Zhengzhou, Henan Province, P. R. China

Key Laboratory of Precision Diagnosis and Treatment for Chronic Kidney Disease in Henan Province, Zhengzhou, P. R. China

Core Unit of National Clinical Medical Research Center of Kidney Disease, Zhengzhou, P. R. China

Yi Li Key Laboratory of Integrated Regulation and Resource Development on Shallow Lakes, Ministry of Education, College of Environment, Hohai University, Nanjing, People's Republic of China

Paula Mantecón Dpto. de Ingenierías Química y Biomolecular, Universidad de Cantabria, Santander, Cantabria, Spain

Bohdana Markiv Dpto. de Ingenierías Química y Biomolecular, Universidad de Cantabria, Santander, Cantabria, Spain

Ana Lourdes Oropesa Toxicology Area, University of Extremadura, Caceres, Spain

BRILiC – Biotechnology Research Institute in Livestock and Cynegetic, University of Extremadura, Caceres, Spain

Laura Ruiz-Azcona Global Health Research Group, Dpto. Enfermería, Universidad de Cantabria-IDIVAL, Santander, Cantabria, Spain

Miguel Santibáñez Global Health Research Group, Dpto. Enfermería, Universidad de Cantabria-IDIVAL, Santander, Cantabria, Spain

Amadeu M. V. M. Soares Biology Department and CESAM, University of Aveiro, Aveiro, Portugal

Godfred Kwesi Teye Key Laboratory of Integrated Regulation and Resource Development on Shallow Lakes, Ministry of Education, College of Environment, Hohai University, Nanjing, People's Republic of China

Yilin Zhang Department of Nephrology, The First Affiliated Hospital of Zhengzhou University, Zhengzhou, P. R. China

Research Institute of Nephrology, Zhengzhou University, Zhengzhou, P. R. China

Research Center for Kidney Disease, Zhengzhou, Henan Province, P. R. China

Key Laboratory of Precision Diagnosis and Treatment for Chronic Kidney Disease in Henan Province, Zhengzhou, P. R. China

Core Unit of National Clinical Medical Research Center of Kidney Disease, Zhengzhou, P. R. China

Investigating the Potential Toxicity of Hydraulic Fracturing Flowback and Produced Water Spills to Aquatic Animals in Freshwater Environments: A North American Perspective

Erik J. Folkerts, Greg G. Goss, and Tamzin A. Blewett

Contents

Abstract Unconventional methods of oil and natural gas extraction have been a growing part of North America's energy sector for the past 20–30 years. Technologies such as horizontal hydraulic fracturing have facilitated the exploitation of geologic reserves that were previously resistant to standard drilling approaches. However, the environmental risks associated with hydraulic fracturing are relatively understudied. One such hazard is the wastewater by-product of hydraulic fracturing processes: flowback and produced water (FPW). During FPW production, transport, and storage, there are many potential pathways for environmental exposure. In the current review, toxicological hazards associated

Electronic supplementary material The online version of this chapter (https://doi.org/10.1007/398_2020_43) contains supplementary material, which is available to authorized users.

E. J. Folkerts (✉) · G. G. Goss · T. A. Blewett
University of Alberta, Department of Biological Sciences, Edmonton, AB, Canada
e-mail: efolkert@ualberta.ca; greg.goss@ualberta.ca; tamzin@ualberta.ca

P. de Voogt (ed.), *Reviews of Environmental Contamination and Toxicology Volume 254*,
Reviews of Environmental Contamination and Toxicology 254,
https://doi.org/10.1007/398_2020_43

with FPW surface water contamination events and potential effects on freshwater biota are assessed. This review contains an extensive survey of chemicals commonly associated with FPW samples from shale formations across North America and median 50% lethal concentration values (LC_{50}) of corresponding chemicals for many freshwater organisms. We identify the characteristics of FPW which may have the greatest potential to be drivers of toxicity to freshwater organisms. Notably, components associated with salinity, the organic fraction, and metal species are reviewed. Additionally, we examine the current state of FPW production in North America and identify the most significant obstacles impeding proper risk assessment development when environmental contamination events of this wastewater occur. Findings within this study will serve to catalyze further work on areas currently lacking in FPW research, including expanded whole effluent testing, repeated and chronic FPW exposure studies, and toxicity identification evaluations.

Keywords Formations · Hazards · Invertebrates · LC_{50} · Major ions · Metals · Natural gas · Oil · Organics · PAH · Regulation · Salts · Spills · Trace metals · Vertebrates · VOC · Wastewater

Abbreviations

AB	Alberta
AER	Alberta Energy Regulator
Ag	Silver
AhR	Aryl hydrocarbon receptor
Ba	Barium
BC	British Columbia
BCOGC	British Columbia Oil and Gas Commission
BTEX	Benzene, toluene, ethylbenzene, and xylene
Ca	Calcium
Cd	Cadmium
Cu	Copper
DBCM	Dibromochloromethane
DOC	Dissolved organic carbon
ER	Estrogen receptor
Fe	Iron
FPW	Flowback and produced water
HCB	Hexachlorobenzene
HPF	Hours post fertilization
HPG/I	Hypothalamic-pituitary-gonadal/interrenal
ILCM	Interlamellar cell mass
K	Potassium
KOW	Octanol-water partitioning coefficient
LC_{50}	Lethal concentration causing 50% Mortality
Mg	Magnesium

Mn	Manganese
MYA	Million years ago
Na	Sodium
Ni	Nickel
NOS	Nitric oxide synthase
PAH	Polycyclic aromatic hydrocarbon
Pb	Lead
PCP	Pentachlorophenol
PXR	Pregnane X receptor
ROS	Reactive oxygen species
Sr	Strontium
TDS	Total dissolved solid
Tl	Thallium
UOG	Unconventional oil and gas
VOC	Volatile organic compound
WET	Whole effluent toxicity
Zn	Zinc

1 Introduction

Extraction of hydrocarbons from unconventional resources is increasing globally as a result of technological advances in horizontal drilling (Gagnon et al. 2016; Vengosh et al. 2014). Hydraulic fracturing is one such advance where chemically formulated fluids are used to fracture low permeability formations under high pressure and temperature (Stringfellow et al. 2014). This process increases the permeability of shale, tight sands, coalbeds, and other gas and oil containing strata, facilitating extraction of hydrocarbon resources from reserves that would otherwise be uneconomic (Stringfellow et al. 2014). Hydraulic fracturing fluids are injected over a short period of time and use large quantities of water per well injection (10,000–100,000 m^3). Injected fluids are comprised of chemical additives that maximize efficiency and output of oil and gas from any given well. These additives include proppants (e.g., ceramic beads, sand – to prevent fracture reclosure), biocides (to prevent microbial degradation of end product resources and maintain well viability), gelling and foaming agents, pH adjustors, clay stabilizers, and surfactants (Ferrer and Thurman 2015; DiGiulio and Jackson 2016; Lester et al. 2015; FracFocus.org). When pressure along the length of the fracture is released, these complex chemical mixtures of hydraulic fracturing fluid return to the surface, and this fluid is termed either flowback or produced water. The delineation of "flowback" versus "produced water" is often subject to well operator discretion. Flowback typically is considered the earliest fluid which returns to the surface and which most closely resembles the composition of the initial injection fluid.

Produced water is the fluid returning from the subterranean environment following longer periods of well production. This fluid is often saline in nature and contains high gas and oil content, which is then operationally separated (U.S. EPA 2016). The distinction between these two types of fluid is not clear, since mixing occurs in the formation, so for the purposes of this review, we will refer to the hydraulic fracturing wastewater as flowback and produced water (FPW) (Stringfellow et al. 2014) with the recognition that toxicological properties of FPW change over the course of flowback production. FPW is a chemically complex heterogeneous mixture that contains highly variable concentrations of organic compounds (e.g., polycyclic aromatic hydrocarbons, PAHs), naturally occurring radioisotopes (e.g., radium), ions (e.g., calcium, magnesium, chloride, sodium, potassium), metals (e.g., barium, thallium, lead), and transformation products resulting from chemical reactions under high temperature and high pressure (DiGiulio and Jackson 2016; He et al. 2017a; Lester et al. 2015) (further expansion in Sect. 2). The specific composition of any given FPW is unique, depending on many factors including the geology of the formation, well shut-in length, phase of flowback collection, and the composition of the initial fracturing fluid additives used for operation (Alessi et al. 2017; Goss et al. 2015; Stringfellow et al. 2017). FPW may be recycled but must eventually be disposed of.

It is estimated that in the USA, unconventional production of natural gas will account for nearly half of newly developed gas production by 2035 (Gagnon et al. 2016). Considering this expected rise in unconventional oil and gas (UOG) activity and the large quantities of FPW created from these processes, management strategies for FPW and potential hazards to the environment are two growing concerns facing the industry and regional governments. One of the key risks associated with hydraulic fracturing is the potential for spills during the generation and transport of large volumes of FPW. Such risks include ground and surface water contamination during pipeline leaks, truck transportation, and injection well integrity issues (Ferrar et al. 2013). Determination of ecological impacts of FPW is considered a top science priority to inform energy policy, conservation, and management of natural systems (Jones et al. 2015). As such, government, industry, landowners, and environmental groups have scrutinized the current methods of storage, transportation, and remediation protocols when spills/leaks occur of this wastewater (Boudet et al. 2014; Gehman et al. 2016; Theodori et al. 2014).

Despite rigorous regulation and controls in all North American jurisdictions on the handling of FPW, spills still occur. Depending on the state/province, a spill or release of FPW warranting report can vary in terms of volume of spill, spill type, and time required to report a spill. For example, in both North Dakota and Colorado, spills escaping secondary containment of $\geq$42 gal must be reported, with written report of said spills mandated within at least 10 days (Patterson et al. 2017). In Pennsylvania, however, spills of $\geq$5–15 gal (depending on FPW TDS measurements) are required within 24 h, whereas in New Mexico, only spills $\geq$210 gal require written notification within 15 days (Patterson et al. 2017). In the province of Alberta, Canada, the Alberta Energy Regulator (the main energy policy and regulating agency in the province) requires spills of $\geq$2000 L (~ 530 gal) to be reported within 24 h (Alberta Energy Regulator n.d.; AER). Regarding spill frequency, in Alberta, an estimated >2,500 FPW spills occurred from 2005 to 2012 with more than 113 of those spills entering into freshwater lakes and streams (Alessi et al. 2017; Goss et al. 2015). Analysis of data from 2005 to 2014 representing >30,000 UOG wells in the states of Colorado, North Dakota, Pennsylvania, and New Mexico showed that 50% of spills were associated with storage or transport of FPW and that 2–16% of wells reported a spill every year, the largest singular FPW spill recorded at 3,756 m^3 (Patterson et al. 2017). However, in this same study, it was found overall that across the three states of Colorado, North Dakota, and Pennsylvania, annual spill rates are either being sustained or decreasing, with only New Mexico showing an increasing annual spill rate (Patterson et al. 2017). With increased use of UOG technologies on the rise recently, however, it is only expected that spills of FPW will continue and become more prevalent. Furthermore, risks associated with FPW are not only from spills but from direct application for dust suppression or de-icing on roads. Currently, 13 states in the USA allow spreading of FPW on roads as an inexpensive alternative to other dust suppressants (Tasker et al. 2018). FPW released to the environment in this manner has the potential to leach from roads during rain events into ground and surface water and cause toxicity to aquatic biota. In Northern Pennsylvania alone, over 280,000 L of FPW was spread on roads in 2015 (Pennsylvania Department of Environmental Protection 2016) resulting in a mean radium concentration of 14.5 pCi/L in surface waters associated with the contaminated roadways, significantly higher than national recommended regulatory limits for radium in drinking water (5 pCi/L) (Tasker et al. 2018).

This review summarizes the known chemical composition of FPW and characterizes the risks to the aquatic environment associated with FPW spills. We have collected and collated FPW characterization data from multiple independent published academic, government, and industry sources on North American formations used for UOG practices and identify three largest and most commonly observed components of FPW which pose the greatest risk to aquatic animals in a spill event. These components include organics, trace metals, and major ions. After identifying these common components, the toxicity of specific chemicals pertaining to these components is further investigated by reviewing respective published lethal and sublethal effects for each chemical. The few studies performed to date which have specifically examined the toxicity of FPW (or simulated FPW) are also reviewed and used to identify the progress of FPW research on freshwater biota and to highlight specific areas requiring further investigation.

2 FPW Chemical Characterization

In the current review, three major classes of chemical species were taken into consideration when analyzing the composition of FPW samples: major ions and other general water characteristics, organic constituents, and trace metals. Although radioactivity associated with FPW has also been determined to be of potential concern to natural environments (Haluszczak et al. 2013; Tasker et al. 2018), analysis of FPW radioactivity was not included in this review. However, a table recording types and levels of radioactivity associated with FPW can be found in the Supplemental Information (SI Table 1). Furthermore, we have restricted our analyses to only those constituents that are commonly reported across studies of FPW composition from North American hydraulic fracturing operations. In total, 15 different sources of detailed chemical analysis of strictly FPW (not affected surface waters or effluent discharges) were compiled to create a single database of over 5,000 data points of targeted inorganic and organic chemicals. These sources can be found in Table 1. It should be noted that the analyzed FPW samples detailed in the literature vary with respect to the timing of collection (flowback period sampled), well location, and shale formation being exploited – all factors which influence FPW chemistry. Indeed, many reports do not offer sufficient or complete details regarding these variables. Consequently, it is therefore acknowledged that the current review is not able to offer detail regarding how the chemical composition and toxicity of FPW samples differ with respect to time spent in the well or source geology. For information on hydraulic fracturing-associated wastewaters from drilling operations in other non-North American formations and on potential other toxicities pertaining to the initial fracturing fluid used to induce fractures, please see reviews by Faber et al. (2017) and Annevelink et al. (2016).

2.1 Major Ions

Cations and anions/salt-related ions (see Table 2) are major components of FPW and are primarily responsible for the high total dissolved solid (TDS) concentrations that are often observed in FPW samples. During the fracturing process, many classes of chemical compounds present in the initial fracturing fluid pumped down the wellbore (e.g., acids, breakers, and stabilizers) contain ionic chemical species that enhance operational efficiency and maximize resource recovery (FracFocus.org). These include chemicals such as hydrochloric acid (used to dissolve minerals and initiate fissure formation), sodium and calcium chloride (breakers used to stabilize geogenic products and clay formations), magnesium peroxide (a breaker used to delay gel break down), and many others. These ions, following parent compound reactions, reside in the formation fluid until they are brought back up to the surface in FPW.

However, another source of salt ionic species contributing to the commonly seen high FPW TDS concentrations is the geological environment being targeted/exploited during hydraulic fracturing activities. During the mid-late Cretaceous era (~100 million years ago; MYA), North America was divided into two land masses by a large inland sea which stretched from the Gulf of Mexico to the Arctic Ocean (Nicholls and Russell 1990). Correspondingly, many of the sedimentary formations exploited for hydraulic fracturing purposes have a marine origin,

Table 1 References used for compiling chemical characterization data

Reference	Type of document	Analytical methods and QA/QC stated?	Location/ formation of FPW sampled	Time period of FPW sample(s) (days post-well stimulation)
US EPA Atlas response_211 419 (2015)	Analytical study report	Yes	Unknown (Marcellus suspected)	Unknown
Blauch et al. (2009)	Regional meeting paper	No	Pennsylvania/ Marcellus	0–55
Blewett et al. (2017a)	Academic paper	Yes	Alberta/ Duvernay	10
DEP_TENORM (2016)	Analytical study report	Yes	Pennsylvania/ Marcellus	Unknown
Dresel and Rose (2010)	Geological survey report	Yes	Pennsylvania/ Marcellus	Unknown
Hayes (2009)	Analytical study report	Yes	Pennsylvania/ Marcellus	0, 1, 5, 14, and 90
He et al. (2017a)	Academic paper	Yes	Alberta/ Duvernay	7
Lauer et al. (2016)	Academic paper	Yes/No	North Dakota/ Bakken	Unknown
Lester et al. (2015)	Academic paper	Yes	Colorado/Denver-Julesburg Basin	Unknown
Maguire-Boyle and Barron (2014)	Academic paper	Yes	Pennsylvania/ Marcellus Texas/Eagle Ford Barnett/New Mexico	Unknown
NYS_DEC (2015)	Environmental impact statement proceeding	Yes/No	Pennsylvania – West Virginia/ Marcellus	Unknown
DEP_Inorganics Report (2010)	Analytical study report	No	Unknown (Marcellus suspected)	Unknown
Rosenblum et al. (2017a)	Academic paper	Yes	Colorado/ Niobrara	1, 4, 7, 15, 22, 80, 130, 220, and 405
Rosenblum et al. (2017b)	Academic paper	Yes	Colorado/ Niobrara	1, 4, 7, 15, 22, 55, 80, 130, and 220
Ziemkiewicz et al. (2014)	Academic paper	Yes	Pennsylvania/ Marcellus	Unknown

Table 2 General water quality and ion concentrations in hydraulic fracturing flowback and produced water samples and current acute (and chronic) guideline concentrations for the protection of freshwater aquatic life according to Canadian Environmental Quality Guidelines (CEQ) and the US Environmental Protection Agency (USEPA)

Chemical	$\overline{X}$ conc.	Median conc.	Range	*N*	CEQ guidelines		USEPA guidelines	
					Acute	Chronic	Acute	Chronic
Bromide	610	472	0.2–2,240	147	ND	ND	ND	ND
Calcium	9,252	7,630	35.2–41,600	164	ND	ND	ND	ND
Chloride	60,293	43,800	64.2–207,000	193	640	120	860	230
Magnesium	982	710	7.1–13,000	164	ND	ND	ND	ND
Manganese	9	4	0–96.5	160	ND	ND	ND	ND
Potassium	705	253	2.7–17,043	148	ND	ND	ND	ND
Sodium	26,669	21,510	45.9–95,500	165	ND	ND	ND	ND
Sulfate	94	46	0–1,010	167	ND	ND	ND	ND
TDS	108,562	91,405	680–345,000	148	ND	ND	ND	ND
Nitrate	2.049	1.3	0.081–15.9	74	550	13	ND	ND
Nitrite	12.03	4.7	0.045–146	72	ND	60	ND	ND
Nitrogen (total)	116	96	5.6–498	69	ND	ND	ND	ND
Carbon (total)[a]	1,200	110	1.2–58,550	77	ND	ND	ND	ND
pH	6.3	6.5	3.4–10.1	126	ND	6.5–9.0	ND	6.5–9.0

ND no data, *N* number of collective samples used to establish range. All values in mg/L
[a]Carbon (total) includes measurements of total organic carbon, total carbon, and dissolved organic carbon

and as a consequence, formation waters and FPW from these formations following natural resource production contain high levels of major ions/salts (Connolly et al. 1990; Kahrilas et al. 2016; Li et al. 1997; Rice 2003; Zhuoheng and Osadetz 2013).

2.2 *Trace Metals*

FPW metal profiles depend greatly on the geology of the formation being exploited. As seen in Table 3, numerous different trace metals may be present in FPW and at varying concentrations. The trace metals present in FPW which will be discussed in this section are from multiple different chemical groups, including the alkali and alkaline earth metals (e.g., barium and strontium – both of which are commonly found at high concentrations in FPW), classically defined "transition" metals (e.g., iron, zinc, cadmium, etc.), post-"transition" metals (e.g., aluminum, lead, thallium, etc.), and metalloids (e.g., boron, arsenic, etc.). Due to their toxicity at relatively low concentrations (Wood 2012a), many of these metals may pose significant hazards to aquatic systems if present (detailed in Sect. 3.2).

Table 3 Trace metal species present in hydraulic fracturing flowback and produced water samples and current acute (and chronic) guideline concentrations for the protection of freshwater aquatic life according to Canadian Environmental Quality Guidelines (CEQ) and the US Environmental Protection Agency (USEPA)

					CEQ guidelines		USEPA guidelines	
Chemical	$\overline{X}$ conc.	Median conc.	Range	*N*	Acute	Chronic	Acute	Chronic
Antimony	0.25	0.1	0–9.9	71	ND	ND	ND	ND
Arsenic	0.37	0.077	0–26.1	93	ND	0.005	0.34	0.15
Barium	1,177.32	355	0–13,900	157	ND	ND	ND	ND
Beryllium	0.031	0.04	0–0.1	89	ND	ND	ND	ND
Cadmium	0.024	0.01	0–0.13	93	0.001	9.0×10^{-5}	0.0018	7.2×10^{-4}
Copper	0.21	0.071	0–4.15	125	ND	0.002–0.004[a]	ND	0.0065–0.034[a,b]
Iron (total)	203	46	0–18,432	163	ND	0.3	ND	1.0
Lead	0.11	0.03	0–3.48	118	ND	0.001–0.007[a]	0.082	0.0032[a]
Mercury	0.18	0.0002	0–14.6	83	ND	2.6×10^{-5}	0.0014	7.7×10^{-4}
Strontium	1,632.73	912	0.58–13,100	160	ND	ND	ND	ND
Thallium	2.34	0.1	0.0049–151	69	ND	8.0×10^{-4}	ND	ND
Zinc	8	0.1	0–685	128	0.037	0.007	0.12	0.12

ND No Data, *N* number of collective samples used to establish range. All values in mg/L
[a]Concentrations dependent on water hardness
[b]Concentrations based on a biotic ligand model analyses

2.3 Organic Chemicals

Similar to major ions, organic chemicals present in FPW are derived from two primary sources: from chemicals originally added to fracturing fluids for the purposes of inducing formation fractures and maintaining well viability but also from the formations themselves. Organic chemicals present in FPW samples derived from the formations being exploited are petrogenic in nature. As noted previously, certain drilling practices add specific organic chemicals to the fracturing fluid to aid extraction (e.g., biocides for antifouling properties and gelling agents such as ethylene glycols and differing petroleum distillates). Many of these fracturing fluid additives and their purposes can be found on the open-access website FracFocus.org. Although the majority of fracturing fluid organic additives are depleted during drilling operations, trace amounts of organic additives may still be present in collected FPW samples. However, much of the reported data on FPW does not include analysis of these organic additives. Thus, for the sake and purpose of this review, only organic chemicals originating from the formations that are present in FPW will be discussed.

The organic species present in FPW are both numerous and diverse. Although only organic chemicals that are common to most FPW samples are reported in Table 4, potentially thousands of organic chemical species may be present (He et al. 2017a). One of the more commonly measured organic chemicals in FPW are polycyclic aromatic hydrocarbons (PAHs), a class of organic compounds that have high toxicity, and which elicit negative impacts on both biota and ecosystems through many different mechanisms. This group encompasses both small organic chemicals, such as the double to 4-ring PAHs (e.g., naphthalene, phenanthrene, and pyrene), and relatively larger 5 to 10-ring PAH molecules (e.g., benzo[a]pyrene, perylene, and ovalene). It should be noted, however, that many larger PAH compounds ($\geq$6 rings) are often insoluble in water and sorb to organic carbon in aquatic systems (de Maag et al. 1998; Ma et al. 2010). Accordingly, such larger PAHs are immediately less bioavailable to aquatic organisms (although accumulation within sediments may become a toxicological concern).

2.4 Wellbore Reaction Products

One area of FPW chemical characterization which has recently begun to gain analytical attention is the presence of transformation products created during reactions involving the initial fracturing fluid compounds and the high-heat, high-pressure environments of a horizontally fractured wellbore. Albeit highly variable, and dependent on numerous factors including depth of formation, length of well, formation geology, etc., temperatures and pressures within a bore of a horizontally fractured well can reach up to ~200°C and 10,000 psi, respectively (Kahrilas et al. 2016; Nelson and Santus 2011; Shaffer et al. 2013). This creates

Table 4 Organic chemical species present in hydraulic fracturing flowback and produced water samples and current acute (and chronic) guideline concentrations for the protection of freshwater aquatic life according to Canadian Environmental Quality Guidelines (CEQ) and the US Environmental Protection Agency (USEPA)

Chemical	$\overline{X}$ conc.	Median conc.	Range	*N*	CEQ guidelines		USEPA guidelines	
					Acute	Chronic	Acute	Chronic
Polycyclic aromatic hydrocarbons								
Naphthalene	68.66	5	0–1,400	69	ND	1.1	ND	ND
Acenaphthylene	8.084	1.9	0–190	69	ND	ND	ND	ND
Acenaphthene	8.34	1.9	0–190	69	ND	5.8	ND	ND
Fluorene	7.89	1.9	0–190	69	ND	3	ND	ND
Phenanthrene	7.64	1.9	0.15–190	69	ND	0.4	ND	ND
Fluoranthene	7.67	1.9	0–190	69	ND	0.04	ND	ND
Pyrene	7.71	1.9	0.037–190	69	ND	0.025	ND	ND
Benzo[a]anthracene	8.085	1.9	0–190	69	ND	0.018	ND	ND
Chrysene	8.026	1.9	0–190	69	ND	ND	ND	ND
Benzo[b]fluoranthene	8.21	1.9	0–190	69	ND	ND	ND	ND
Benzo[k+j]fluoranthene	8.15	1.9	0–190	69	ND	ND	ND	ND
Benzo[a]pyrene	8.18	1.9	0–190	69	ND	0.015	ND	ND
Indeno[1,2,3-cd]pyrene	8.25	1.9	0–190	69	ND	ND	ND	ND
Benzo[g,h,i]perylene	7.89	1.9	0–190	69	ND	ND	ND	ND
Dibenz[a,h]anthracene	8.27	1.9	0–190	69	ND	ND	ND	ND
(1 or 2)-Methylnaphthalene	11.19	1.9	0.11–190	69	ND	ND	ND	ND
Volatiles								
1,1,1-Trichloroethane	46.1	5	0–500	68	ND	ND	ND	ND
1,1,2-Trichloroethane	46.1	5	0–500	68	ND	21	ND	ND
1,2-Dichloropropane	46.1	5	0–500	68	ND	ND	ND	ND
1,2,4-Trichlorobenzene	43.35	5	0–500	68	ND	24	ND	ND
Benzene	888.6	7.7	0–12,500	81	ND	370	ND	ND

(continued)

Table 4 (continued)

Chemical	$\overline{X}$ conc.	Median conc.	Range	*N*	CEQ guidelines		USEPA guidelines	
					Acute	Chronic	Acute	Chronic
Carbon tetrachloride	46.1	5	0–500	68	ND	13.3	ND	ND
Ethylbenzene	103.9	5	0–1,540	75	ND	90	ND	ND
Styrene	45.4	5	0–500	70	ND	72	ND	ND
Toluene	1,293.3	5	0–41,500	81	ND	2	ND	ND
Vinyl chloride	46.1	5	0–500	68	ND	ND	ND	ND
Xylenes (total)	1,569.3	15	3–73,000	76	ND	ND	ND	ND
Synthetics								
Dibromochloromethane	46.1	5	0–500	68	ND	ND	ND	ND
Hexachlorobenzene	8.21	1.9	0–190	68	ND	ND	ND	ND
Pentachlorophenol	40.98	9.5	0–940	68	ND	0.5	19	15

ND No Data, *N* number of collective samples used to establish range. All values in μg/L

an ideal environment for the transformation of organic chemicals incorporated into the initially injected fluid. These wellbore-created chemical species may hold differing toxicities than the original parent compounds (He et al. 2017a; Kahrilas et al. 2016). Although theorized to constitute a relatively small proportion of the total FPW chemical makeup, these wellbore-transformed compounds pose challenging problems to operators and regulatory agencies when attempting to develop risk assessment and remediation protocols. Currently, targeted analysis of select compounds is slowly elucidating the extent of these transformational reactions and the occurrence of these wellbore reaction products. However, non-targeted approaches (such as non-targeted orbitrap GC-MS studies) comparing FPW and initial fracturing fluids may better determine what and how many products are formed in a given fracturing operation. Industry proprietary constraints, however, often impede these types of studies and hinder further expansion of this topic.

3 FPW Chemical Constituent Toxicity

3.1 Major Ion Toxicity

Salinity in produced waters is likely to be the most significant source of acute toxicity associated with FPW exposure to aquatic organisms (Folkerts et al. 2019). The levels of salts such as Na (sodium), K (potassium), Mg (magnesium), Ca (calcium), and Cl^- (chloride) in many FPWs are extremely high compared to natural freshwaters. For example, the levels of Na in FPW may be >6,000-fold higher than Na concentrations in some freshwater lakes and rivers (Table 2; Allen 2008). As observed with trace metals, there is a large variation within concentrations of these major ions. For example, Na concentrations in FPW can range from 45.9 to 95,599 mg/L, and Cl values range from 64.2 to 207,000 mg/L depending on the FPW sample (Table 2). In assessing the potential for major ion toxicity, it is worth noting that it may be difficult to isolate the effects of specific ions or salts, as elevations of multiple ions co-occur, and this is a phenomenon likely to influence toxicological outcomes. However, as a general metric, salinity can be a useful guide as to the anticipated toxic outcome, and concentrations of major ions in FPW often surpass Canadian Environmental Quality (CEQ) and United States Environmental Protection Agency (USEPA) water quality guideline concentrations and LC_{50} values for numerous aquatic organisms (see Tables 2 and 5). A sudden influx of highly saline water, such as a saline FPW spill, into a lake or stream will initially cause toxicity due to osmoregulatory distress. All freshwater organisms must closely maintain ionic homeostasis, but the species most at risk are stenohaline species that may lack the capacity to osmoregulate beyond their normal environmental salinity range (Evans et al. 2005).

Similar to metals, the gill is a primary target for the effects of salts associated with an FPW spill. The gill is a multifunctional organ that is involved with ion regulation and is the first line of defense against toxicants. Morphologically, the

Table 5 Reported acute lethal toxicities (LC_{50} in mg/L) of major ions associated with hydraulic fracturing flowback and produced water to aquatic species

Chemical	Species	LC_{50} 48 h	LC_{50} 96 h	Water temp. (°C)	Hardness (mg/L $CaCO_3$)	pH	Reference
Bromide	Juv. DM	1.0[a]	–	22	173	8	LeBlanc (1980)[c]
		8.1[a]	–	20	170	7.5	Baird et al. (1991)[c]
		–	>100.0[a]	20	130	7.5	Ewell et al. (1986)[c]
		11,000.0[a]	–	20	–	–	Canton and Wester (1983)[c]
		13,500.0[a]	–	19	100	–	Hermens et al. (1984)[d]
	Juv. FHM	–	>100.0[a]	20	130	7.5	Ewell et al. (1986)[c]
	FHM	17,800.0[a]	16,500.0[a]	12	107	7.8	Alexander et al. (1981)[c]
Calcium	Juv. DM	52.0[a]	–	18	45	7.7	Biesinger and Christensen (1972)[d]
	Juv. DM	2,770.0[a]	–	20	Mod. Hard	7.5–9.0	Mount et al. (1997)[c]
	FHM	–	4,630.0[a]	25	Mod. Hard	7.5–9.0	Mount et al. (1997)[c]
Magnesium	Juv. DM	140.0[a]	–	18	45	7.7	Biesinger and Christensen (1972)[d]
	Juv. DM	1,330.0[a]	–	20	Mod. Hard	7.5–9.0	Mount et al. (1997)[c]
	FHM	–	2,120.0[a]	25	Mod. Hard	7.5–9.0	Mount et al. (1997)[c]
Manganese	Juv. DM	9.8[a]	–	18	45	7.7	Biesinger and Christensen (1972)[d]
Potassium	Juv. DM	93.0[a]	–	18	45	7.7	Biesinger and Christensen (1972)[d]
	Juv. DM	660.0[a]	–	20	Mod. Hard	7.5–9.0	Mount et al. (1997)[c]
	FHM	–	880.0[a]	25	Mod. Hard	7.5–9.0	Mount et al. (1997)[c]
Sodium	Juv. DM	1,640.0[a]	–	18	45	7.7	Biesinger and Christensen (1972)[d]
	Juv. DM	4,770.0[a]	–	20	Mod. Hard	7.5–9.0	Mount et al. (1997)[c]
	FHM	–	6,390.0[a]	25	Mod. Hard	7.5–9.0	Mount et al. (1997)[c]

(continued)

Table 5 (continued)

Chemical	Species	LC_{50}		Water temp. (°C)	Hardness (mg/L $CaCO_3$)	pH	Reference
		48 h	96 h				
Sulfate	CD	2,050.0[a]	–	25	94	7.9	Soucek and Kennedy (2005)[d]
		2,526.0[a]	–	25	107	7.9	Soucek and Kennedy (2005)[d]
	HA	512.0[a]	–	22	94	7.9	Soucek and Kennedy (2005)[d]
		2,855.0[a]	–	22	107	7.9	Soucek and Kennedy (2005)[d]

CD Ceriodaphnia dubia, DM Daphnia magna, FHM fathead minnow (*Pimephales promelas*), *HA Hyalella azteca, Juv.* juvenile-aged organisms used. DM = within 24 h. RT = ≤ 1-year post hatch

[a]Static/semi-static exposures

[b]Flow-through exposures

[c]Nominal chemical concentrations used for LC_{50} calculations

[d]Measured chemical concentrations used for LC_{50} calculations

gill epithelium is composed of several different cell types, which undergo changes in response to physiological stress and environmental challenges. Of these cell types, interlamellar cells (cells between the lamellae of gills) may grow and change to meet physiological demands, such as in situations where environmental dissolved oxygen concentrations are altered (Sollid and Nilsson 2006), toxicant exposure occurs (Blewett et al. 2017b), or when the fish is osmotically challenged (Blair et al. 2016). Changes to interlamellar cell mass (ILCM) is a commonly used index for observing respiratory stress in fish (Sollid and Nilsson 2006). Previous studies have observed altered responses of the ILCM as a response to high salt exposure (Blair et al. 2016; Blewett et al. 2017c). In Arctic grayling (*Thymallus arcticus*) and rainbow trout (*Oncorhynchus mykiss*) exposed to salinities of 17,000 and 13,500 mg/L, respectively, for 24 and 48 h, histological analysis of the gill filaments showed structural changes. In the stenohaline Arctic grayling, ILCM recordings showed signs of hyperplasia and overgrowth of the protruding lamellae, decreasing the surface area of the gills dramatically. This gill remodeling is thought to be beneficial as a defense mechanism against exposure to high saline FPW as it would limit the uptake of salts and minimize water loss across the epithelium. Conversely, during the same exposure, the ILCM was reduced in rainbow trout, likely because of the greater osmoregulatory capacity of this euryhaline species was sufficient to handle the change in environmental concentrations.

Another acute effect observed in fish exposed to high saline waters is the generation of an oxidative stress response. In trout exposed to the salinity component of FPW alone, alterations in key oxidative stress enzyme activities were observed in the gill and liver of rainbow trout (Blewett et al. 2017c). More specifically, there was both an increase in the enzyme catalase activity and a decrease in the enzyme superoxide dismutase activity. These enzymes are part of a critical antioxidant defense system which operates to scavenge free oxygen radicals, preventing damage to cellular constituents. For an extensive review on salinity-induced oxidative stress, see Lushchak (2011). *Daphnia magna*, a freshwater crustacean, showed even more extreme sensitivity to the major ions present in FPW (Blewett et al. 2017a). In this study, the authors exposed the animals to dilutions of a saline solution which was made up to match the major ion composition of a specific FPW sample. The median lethal concentration occurred when the saline solution was diluted to a concentration representing just 1.5% (~2,700 mg/L) of the raw FPW sample salinity. Previous reports have shown that total offspring production in *Daphnia* decreases under NaCl exposure (Ghazy et al. 2009; Martínez-Jerónimo and Martínez-Jerónimo 2007). Blewett et al. (2017a) showed that after chronic exposure, both salinity and FPW caused decreased reproduction and metabolic rate. Furthermore, chronic exposure to salinity can cause changes in physiological homeostasis such as plasma-enhanced stress-related hormones (e.g., prostaglandins), stimulation of energy metabolism, alteration of electrolyte equilibrium, enhanced reactive oxygen species generation, and altered reproductive fitness (Lushchak 2011).

3.2 *Trace Metal Toxicity*

It is difficult to quantify the toxicity caused by individual trace metals present in FPW, as they are present in mixtures with other trace metals and chemical additives, which modify bioavailability, bioaccumulation, and toxicity. Below, we identify certain trace metals of toxicological concern commonly found in characterized FPW samples by comparing concentrations to LC_{50} values for multiple model aquatic animal species. Furthermore, we outline potential mechanisms of toxicity of individual trace metals to aquatic organisms and the role of metal speciation in mediating toxicity. Most common trace metals found within many different FPW samples studied include barium (Ba), strontium (Sr), iron (Fe), manganese (Mn), thallium (Tl), zinc (Zn), copper (Cu), and lead (Pb). Although these metals are commonly found in FPW samples, their concentrations will vary considerably as a consequence of the geology of the formation being exploited for resources. Nevertheless, water quality criteria for dissolved metals are typically in the μg/L range (Wood 2012a), while trace metals present in raw (i.e., undiluted) FPW samples are present at concentrations in the mg/L range. Moreover, FPW frequently reaches metal concentrations well above both the current ambient water quality criteria for Canada, the USA, and the European Union (Wood 2012a) (Table 3) and above LC_{50} concentrations for numerous aquatic organisms (Table 6).

Metal toxicity can be exerted over either an acute time frame or a chronic time frame, and the mechanisms of effect may differ between these two periods

Table 6 Reported acute lethal toxicities (LC_{50} in mg/L) of metals associated with hydraulic fracturing flowback and produced water to aquatic species

Chemical	Species	LC_{50}		Water temp. (°C)	Hardness (mg/L $CaCO_3$)	pH	Reference
		48 h	96 h				
Antimony	DM	>530.0[a]	–	22	173	8	LeBlanc (1980)[b]
Antimony ($^{+3}$)	Juv. PM	15.5[a]	12.4[a]	20	Seawater	~6.5	Takayanagi (2001)[c]
Antimony ($^{+4}$)	Juv. PM	0.93[a]	0.93[a]	20	Seawater	8	Takayanagi (2001)[c]
Arsenic	Juv. DM	2.5[a]	–	21	Modified M4 medium		Tišler and Zagorc (2002)[c]
		7.4[a]	–	18	45	7.7	Biesinger and Christensen (1972)[c]
	Juv. RT	23.2[a]	15.3[a]	12	250	8	Tišler and Zagorc (2002)[c]
Barium	Juv. DM	14.5[a]	–	18	45	7.7	Biesinger and Christensen (1972)[c]
	DM	410[a]	–	22	173	8	LeBlanc (1980)[b]
Beryllium	Juv. DM	–	0.4[a]	20	130	7.5	Ewell et al. (1986)[b]
	DM	1.0[a]	–	22	173	8	LeBlanc (1980)[b]
	Juv. FHM	–	>100.0[a]	20	130	7.5	Ewell et al. (1986)[b]
Cadmium	Juv. DM	0.03[a]	–	19	100	–	Canton and Slooff (1982)[c]
		0.046[a]	–	19	–	7.9	Slooff et al. (1983)[b]
	DM	0.035[a]	–	20	29	7.8	Nelson et al. (1984)
	ZF	7.0[d]	–	20	170	–	Canton and Slooff (1982)[c]
	Juv. FHM	2.2[a]	–	20	140	–	Slooff et al. (1983)[b]
	Juv. RT	–	<0.0005[a]	15	9.2	7	Cusimano et al. (1986)[c]
		–	0.0025[d]	13	20	7.2	Hollis et al. (2000)[c]
		–	0.018[d]	15	140	8	Szebedinszky and McGeer (2001)[c]
		0.091[a]	–	15	20	7.2	Calamari et al. (1980)[c]
		0.15[a]	–	15	285	7.5	Slooff et al. (1983)[b]
		0.36[a]	–	15	80	7.2	Calamari et al. (1980)[c]
	RT	–	0.022[d]	14	140	8	–

(continued)

Table 6 (continued)

Chemical	Species	LC_{50} 48 h	LC_{50} 96 h	Water temp. (°C)	Hardness (mg/L $CaCO_3$)	pH	Reference
Copper	Juv. DM	0.0098[a]	–	18	45	7.7	Biesinger and Christensen (1972)[c]
	DM	0.017[a]	–	20	250	7	Park et al. (2009)[b]
	DP	0.011[a]	–	20	37	7.4	Boeckman and Bidwell (2006)[c]
	Larval FHM	–	0.078[d]	22	–	8.1	Erickson et al. (1996)[c]
		–	0.11[a]	22	–	7.9	Erickson et al. (1996)[c]
	FHM	–	0.075[d]	25	31	7	Mount and Stephan (1969)[c]
		–	0.43[a]	23	200	8	Mount (1968)[c]
		–	0.47[a]	22	~170	–	Dwyer et al. (2005)[b]
	Juv. RT	–	0.0028[a]	15	9.2	7	Cusimano et al. (1986)[c]
		–	0.1[d]	14	120	8	Taylor et al. (2000)[b]
	RT	–	0.03[a]	15	30	9	Howarth and Sprague (1978)[c]
		–	0.031[a]	15	102	7	Howarth and Sprague (1978)[c]
		–	0.3[a]	15	361	7	Howarth and Sprague (1978)[c]
Iron ($^{+3}$)	Juv. DM	9.6[a]	–	18	45	7.7	Biesinger and Christensen (1972)[c]
Lead	Juv. DM	0.45[a]	–	18	45	7.7	Biesinger and Christensen (1972)[c]
	Juv. RT	–	1.0[a]	11	120	8	Rogers et al. (2003)[c]
	RT	–	1.17[a]	11	28	7	Davies et al. (1976)[c]
Mercury	Juv. DM	0.003[a]	–	20	–	7.9	Canton and Adema (1978)[b]
		0.005[a]	–	19	–	7.9	Slooff et al. (1983)[b]
		0.005[a]	–	18	45	7.7	Biesinger and Christensen (1972)[c]
	Juv. FHM	0.05[a]	–	20	140	–	Slooff et al. (1983)[b]
	FHM	–	0.17[a]	23.5	45	7.4	Snarski and Olson (1982)[c]
	Juv. RT	0.65[a]	–	15	285	7.5	Slooff et al. (1983)[b]
Strontium	Juv. DM	125.0[a]	–	18	45	7.7	Biesinger and Christensen (1972)[c]
Thallium	Juv. DM	2.2[a]	–	22	173	8	LeBlanc (1980)[b]

(continued)

Table 6 (continued)

Chemical	Species	LC_{50} 48 h	LC_{50} 96 h	Water temp. (°C)	Hardness (mg/L $CaCO_3$)	pH	Reference
Zinc	Juv. DM	0.1[a]	–	18	45	7.7	Biesinger and Christensen (1972)[c]
		–	0.75[a]	20	130	7.5	Ewell et al. (1986)[b]
	Juv. FHM	–	0.6[d]	25	46	7.5	Benoit and Holcombe (1978)[c]
		–	9.2[d]	23	203	7.7	Brungs (1969)[c]
		–	17.0[a]	20	130	7.5	Ewell et al. (1986)[b]
	Juv. RT	–	0.103[a]	10	10	5.8	Alsop and Wood (1999)[c]
		–	0.17[a]	15	31	7	Bradley and Sprague (1985)[c]
		–	0.869[a]	15	120	8	Alsop et al. (1999)[c]
		–	4.46[a]	15	386	8	Bradley and Sprague (1985)[c]
	RT	–	0.066[a]	15	9.2	7	Cusimano et al. (1986)[c]

DM Daphnia magna, DP Daphnia pulex, FHM fathead minnow (*Pimephales promelas*), *RT* rainbow trout (*Oncorhynchus mykiss*), *ZF* zebrafish (*Brachydanio rerio*), *PM* Red sea bream (*Pagrus major*), *Juv.* juvenile-aged organisms used. DM = within 24 h. RT = $\leq$ 1-year post hatch
[a]Static/semi-static exposures
[b]Nominal chemical concentrations used for LC_{50} calculations
[c]Measured chemical concentrations used for LC_{50} calculations
[d]Flow-through exposures

of exposure. Metals can cause toxicity through a plethora of mechanisms, including structural damage, disruption of enzymatic function, reacting as redox catalysts in the production of ROS, disruption of ion regulation, and initiating the formation of DNA and protein adducts (Liu et al. 2008). Over acute exposure periods, the lethality of metals is often associated with damage at the gill. The gills of fish and invertebrates represent up to 70% of the total surface area of an organism that is directly exposed to the external environment and are organs designed for physiological transport processes (Evans et al. 2005; Hughes 1972). Previous evidence has shown that exposure to various trace metals can cause profound morphological changes in the gills, often characterized by an inflammatory response (Wood 2001). This response can elicit rapid lethality due to edematous swelling, cellular lifting, necrosis, and lamellar fusion – changes that result in a significantly increased water-to-blood diffusion distance, which in turn reduces the ability of the animal to effectively take up oxygen (Wood 2012a).

There are, however, mechanisms of toxicity that are specific to certain trace metals. Some metals impose toxicity via ionic mimicry whereby metals target the active ionic uptake pathways in the gill of aquatic organisms (Bury et al. 2003; Büsselberg 1995; Clarkson 1993; Wood 2012a). For example, silver (Ag) and Cu mimic Na (Goss et al. 2011; Grosell 2012; Grosell and Wood 2002; Wood 2012b), while Pb (Mager 2012), Zn (Hogstrand et al. 1994; Hogstrand 2012), and cadmium (Cd) (McGeer et al. 2012; Niyogi and Wood 2004) will mimic Ca. This mimicry results in these metal contaminants disrupting ion homeostasis, through both decreased uptake of the nutrient ions and through metal effects on sensitive ion transport enzymes. For example, Cu can inhibit the basolateral Na/K ATPase, an enzyme that energizes the active uptake of Na and Cl for ionic regulation (Wood 2012a). Active uptake of Na, Cl, and Ca is essential for freshwater animals, which constantly lose these ions by diffusion from their relatively salt-rich bodies to the much more dilute environment. Consequently, death via hyponatremia or hypocalcemia occurs as a result of metals inhibiting these enzymes. Cu can also inhibit carbonic anhydrase, the enzyme that hydrates CO_2 to produce H^+, which is also critical for acid-based ion regulation (Grosell 2012). Similarly, Zn, Pb, and Cd compete with Ca for entry through apical Ca channels and can inhibit the basolateral Ca-ATPase that powers Ca transport within the gills (Hogstrand 2012; Mager 2012; McGeer et al. 2012).

In terms of chronic toxicity from lower level exposures, similar mechanisms of toxicity may be observed. However, under chronic exposure conditions, a suite of mechanisms are usually altered which ultimately result in decreased survival, growth, or reproduction. For example, small molecular weight thiol-containing proteins such as metallothionein and glutathione may be induced by certain metal exposures and act to bind to intracellular metals and prevent their reaction with sensitive cellular sites (Wood 2012a). However, this may act to limit bioavailability of other essential metals for metalloenzymes. Chronic toxicity of metals may also be associated with increased oxidative stress. Increased metal exposure can induce oxidative stress through two mechanisms, the first is related to the generation of free radicals by ions with changeable valences (e.g., Fe, Ni, Cu) and the second through damage of antioxidant enzymes that are normally used to prevent reactive oxygen species damage (for detailed review of these processes, see Lushchak (2011).

The thresholds for acute and chronic effects of metals will vary as a function of metal concentration and length of exposure. However, in FPW spill scenarios, metals will be present in aquatic settings in the presence of other chemicals that have the potential to mitigate and enhance their toxicity. This occurs either through altering metal speciation or through interactions with effect pathways. The speciation of a trace metal is critical to its toxicity and is significant for understanding trace metal bioavailability and environmental fate (Wood 2012a). Commonly, speciation refers to the distribution of the different metal chemical species within a system. For most metals, the free ion form is considered to be the most toxic, as it is the species that is thought to be the most bioavailable, gaining access to the organism through dedicated ion transporters on epithelial surfaces (as described above; Wood 2012a). In aquatic systems, trace metal bioavailability is affected by complexation (anionic and/or organic) and competition (trace metals and major essential ions [Na, Ca, K, Mg] that may share a pathway; Wood 2012a) making interpretation of the bioavailability in the complex mixture that is FPW extremely challenging.

Anionic complexation involves the association of a trace element with anions such as SO_4^{2-} and Cl^- and OH^- and HCO_3^- (Wood 2012a). The association of any given metal with any given anion is dependent on several factors, including the physicochemical characteristics of both the metal cation and the anion. In FPW, however, concentrations of complexing anions are high. For example, FPW SO_4^{2-} concentrations display a median value of 46.3 mg/L but ranges from 0 to 1,010 mg/L, while the median value for Cl^- is 43,800 mg/L with a range from 64.2 to 207,000 mg/L (Table 2). This suggests that anion complexes of metals are likely to be significant in FPW and in turn presumably affect and alter bioavailability. However, other water chemistry factors play an important role. For example, in freshwater, the chemical speciation of trace metals is greatly influenced by pH. FPW is generally acidic (e.g., median pH of 6.5, Table 2) but can vary greatly from pH 3.4 to 10.1 (Table 2). More alkaline waters will favor bicarbonate or carbonate-based complexes, thereby decreasing bioavailability and toxicity. However, it is important to note that some anion complexes of metals may be bioavailable and induce effects associated with toxicity. For example, the formation of $CuCO_3$ has been associated with the toxic effects of Cu in *Daphnia magna* (De Schamphelaere et al. 2002). However, to date, the toxicological significance of anions and anion complexes has received little attention.

Organometallic complexes are also found in FPW. Metals are known to bind to organic carbon with varying degrees of binding affinities (Aiken et al. 2011; Lores and Pennock 1998; Playle et al. 1993). Again, because of the complexity associated with organics present in hydraulic fracturing effluents, we are unable to individually associate metals with specific organic complexes. However, it is known that PAHs bind to certain metals through cation-π interactions (Gauthier et al. 2014) resulting in lipophilic organometallic complexes. These organometallic complexes may diffuse across the epithelia of aquatic organisms, and the potential for toxic effects associated with this bioaccumulation has been documented (Boullemant et al. 2009; Parthasarathy et al. 2008, 2010). However, to date, the toxicities of these complexes in FPW have not been extensively investigated. However, it must also be noted that metal toxicity in aquatic environments may also be mitigated by complexation with some organics. For example, dissolved organic carbon (DOC) is ubiquitous in natural waters and is formed by the microbial breakdown of plant and animal materials (Thurman 1985). DOC has a high affinity for metals, and by binding to them, bioavailability of trace metals to target surfaces such as the gills is reduced (Wood et al. 2011). Similarly, the high salinity of these effluents (Alessi et al. 2017; Goss et al. 2015) may likely outcompete metals for binding to ion transporters on the gill, reducing uptake, bioavailability, and toxicity (reviewed in Blewett and Leonard 2017).

3.3 Organic Chemicals

In this review, we have categorized organics from analyzed FPW samples into three different subgroups: polycyclic aromatic hydrocarbons, volatiles, and synthetics. Although numerous other organics (and associated groupings) may be present in FPW, analyses performed to date have most readily and reliably identified these main groupings of organic compounds and, thus, will be the focus of discussion in this section.

3.3.1 Polycyclic Aromatic Hydrocarbons

Polycyclic aromatic hydrocarbons (PAHs) have long been identified as a toxicological concern in aquatic environments, with numerous studies investigating the different pathways and processes that PAHs affect and how perturbation of these pathways elicits toxic responses. Our analysis indicates that numerous PAHs exist in FPW samples (Table 4), often at concentrations near or above water quality guidelines for the protection of aquatic life (Table 4) and LC_{50} values for many model aquatic toxicological species (Table 7). As alluded to earlier, many of the PAH compounds present in FPW are petrogenic and derived from the formations being drilled into during horizontal fracturing activities (unless the process uses diesel fuel in the fracturing fluids: Frac Focus Chemical Disclosure Registry n.d.) (Neff et al. 2011; Orem et al. 2014; U.S. EPA 2016). PAHs are generally classified into lower molecular weight (2–3 fused rings) and higher molecular weight (four and above fused rings) compounds with each subgroup having unique physiochemical characteristics associated with their toxicodynamics and toxicity potential. In aquatic systems, lower molecular weight molecules are associated with both the sediment and water matrices, whereas higher molecular weight PAHs generally associate (both physically and chemically) with the sediment (Chiou et al. 1998; Collier et al. 2014; Wang et al. 2001). Degradation via photooxidation and microbial communities occurs for both subcategories of PAHs, although generally, lower molecular weight PAHs have been demonstrated to degrade more rapidly than high molecular weight PAHs, meaning that higher molecular weight PAHs are more environmentally persistent (Bertilsson and Widenfalk 2002; Juhasz and Naidu 2000; Lima et al. 2005; Meador et al. 1995).

Although the presence of both low and high molecular weight PAHs in FPW is readily observable (Table 4), the toxicological consequences of these chemicals in FPW are less well studied. However, for many PAHs, average concentrations found in FPW are at or near LC_{50} values for certain aquatic species (Table 7). Furthermore, for all PAHs identified in FPW samples, the recorded upper-range concentrations are at or above LC_{50} values for almost all species studied. Accounting for the persistent nature of these chemical species, we conclude that a significant PAH hazard exists when FPW is released into an aquatic system.

Table 7 Reported acute lethal toxicities (LC_{50} in mg/L) of polycyclic aromatic hydrocarbons associated with hydraulic fracturing flowback and produced water to aquatic species

Chemical	Species	LC_{50} 48 h	LC_{50} 96 h	Water temp. (°C)	Hardness (mg/L $CaCO_3$)	pH	Reference
Naphthalene	Juv. DM	2.16[a]	–	20	140	7.8	Millemann et al. (1984)[b]
		24.1[a]	–	20	140	7.8	Parkhurst et al. (1981)[b]
	DM	2.19[a,c]	–	20	–	–	Muñoz and Tarazona (1993)[d]
		8.6[a]	–	22	173	8	LeBlanc (1980)[d]
		22.6[a]	–	12	134	7.6	Eastmond et al. (1984)[d]
	Juv. DP	4.66[a,c]	–	20	170	~7.5	Smith et al. (1988)[d]
	DP	–	1.0[a]	15	–	7.5	Trucco et al. (1983)[b]
		3.4[a]	–	20	43	7.2	Geiger and Buikema (1982)[b]
	Juv. FHM	–	1.99[a]	20	140	7.8	Millemann et al. (1984)[b]
	FHM	–	7.9[e]	15	–	–	DeGraeve et al. (1982)[b]
	Juv. RT	–	0.11[e]	13.5	100	7.8	Black et al. (1982)[b]
		–	1.6[e]	15	–	–	DeGraeve et al. (1982)[b]
		–	4.5[a]	12	175	–	Edsall (1991)[d]
Acenaphthene	DM	1.28[a,c]	–	20	–	–	Muñoz and Tarazona (1993)[d]
		41.0[a]	–	22	173	8	LeBlanc (1980)[d]
	Larval FHM	–	0.61[e]	25	35	7.4	Cairns and Nebeker (1982)[c]
	Juv. FHM	–	1.6[e]	23	43	7.5	Holcombe et al. (1983)[b]
	Juv. RT	1.13[e]	0.67[e]	12	46	7.3	Holcombe et al. (1983)[b]
	Larval BT	0.65[e]	0.58[e]	12	46	7.3	Holcombe et al. (1983)[b]
Fluorene	Juv. DP	0.21[a,c]	–	20	170	~7.5	Smith et al. (1988)[d]

(continued)

Table 7 (continued)

Chemical	Species	LC_{50} 48 h	LC_{50} 96 h	Water temp. (°C)	Hardness (mg/L $CaCO_3$)	pH	Reference
Phenanthrene	Juv. DM	0.7[a,c]	–	20	~165	~7.5	Lampi et al. (2006)[d]
		0.7[a]	–	20	140	7.8	Millemann et al. (1984)[b]
		1.0[a]	–	20	140	7.8	Parkhurst et al. (1981)[b]
	DM	0.38[a,c]	–	20	–	–	Muñoz and Tarazona (1993)[d]
		0.84[a]	–	19	134	7.6	Eastmond et al. (1984)[d]
	Juv. DP	0.35[a,c]	–	20	170	~7.5	Smith et al. (1988)[d]
	DP	–	0.1[a]	15	–	~7.5	Trucco et al. (1983)[b]
		1.1[a]	–	20	43	7.2	Geiger and Buikema (1982)[b]
	Juv. RT	–	0.04[e]	13.5	100	7.8	Black et al. (1982)[b]
		–	3.2[a]	12	175	–	Edsall (1991)[d]
Fluoranthene	Juv. DM	0.12[a]	–	22.5	~195	7.8	Spehar et al. (1999)[b]
	DM	0.11[a]	–	20	80	~7	Suedel and Rodgers Jr (1996)[b]
		320.0[a]	–	22	173	8	LeBlanc (1980)[d]
	Larval FHM	–	0.009[e]	25	285	7.1	Diamond et al. (1995)[b]
	Juv. FHM	–	>0.21[a]	22.5	~54	7.8	Spehar et al. (1999)[b]
	Juv. RT	–	0.091[a]	17	~53	7.8	Spehar et al. (1999)[b]
Benz[a] anthracene	Juv. DM	0.0015[a,c]	–	20	165	~7.5	Lampi et al. (2006)[d]
	DP	–	0.01[a]	15	–	7.5	Trucco et al. (1983)[b]

(continued)

Table 7 (continued)

Chemical	Species	LC_{50} 48 h	LC_{50} 96 h	Water temp. (°C)	Hardness (mg/L $CaCO_3$)	pH	Reference
Benzo[a] pyrene	Juv. DM	0.0016[a, c]	–	20	165	~7.5	Lampi et al. (2006)[d]
	DP	–	0.005[a]	15	–	7.5	Trucco et al. (1983)[b]

DM Daphnia magna, DP Daphnia pulex, FHM fathead minnow (*Pimephales promelas*), *RT* rainbow trout (*Oncorhynchus mykiss*), *BT* brown trout (*Salmo trutta*), *Juv.* juvenile-aged organisms used. DM = within 24 h. RT = $\leq$ 1-year post hatch
[a]Static/semi-static exposures
[b]Measured chemical concentrations used for LC_{50} calculations
[c]EC_{50} Measurement
[d]Nominal chemical concentrations used for LC_{50} calculations
[e]Flow-through exposures

PAHs are known to cause numerous effects in exposed organisms. A substantial component of the toxicity associated with PAHs results from both the PAHs and their metabolite intermediates created during organismal biotransformation. For example, intermediate PAH epoxides formed during initial phase I oxidation steps are highly reactive, and these epoxides nonenzymatically form toxic phenol derivatives (Buhler and Williams 1988; Livingstone 1998). Susceptibility to PAH toxicity is highly species dependent, with developmental stage known to be a factor influencing PAH toxicity (James 1989; Meador et al. 1995; Varanasi et al. 1989). When compared to vertebrates, invertebrates exhibit lower rates of both organic contaminant metabolism and elimination of contaminants. As a result, they tend to bioaccumulate metabolizable contaminants such as PAHs to a greater extent (James 1989; Meador et al. 1995). Regardless, biomarkers of PAH biotransformation are still observed in invertebrate species, suggestive of some limited PAH metabolism capacity and potential for PAH intermediate-mediated toxicity (James 1989; Meador et al. 1995; Varanasi et al. 1989). In fish, PAH exposure has been shown to induce hepatic neoplasia and lesions, and many PAHs are known hepatocarcinogens (Baumann et al. 1996; Johnson et al. 1993; Kelly et al. 1993; Pinkney and Harshbarger 2006).

Many vertebrate and invertebrate studies have shown that PAHs negatively impact reproductive processes. In fish, PAHs are known to elicit endocrine-disrupting effects, although the mechanisms behind these PAH effects are notably diverse and complex. One pathway by which PAHs disrupt the endocrine system is through interactions with the aryl hydrocarbon receptor (AhR) and downstream response elements (such as the dioxin response element and the estrogen response element, among others) that regulate endocrine processes (Navas and Segner 2001; Villeneuve et al. 2002; Williams et al. 1998). In addition, PAHs interact with other elements responsible for the expression of several nuclear receptors, including retinoic acid receptors and retinoid X receptors (Bilbao et al. 2010; Cheshenko et al. 2008). Furthermore, PAHs have been demonstrated to influence cytochrome

P450 aromatase activity and expression – a key enzyme involved in the conversion of androgens to estrogens (see reviews Cheshenko et al. 2008; Le Page et al. 2011). PAHs may also influence fish neuronal endocrine processes by altering expression of monoaminergic neurotransmitters such as dopamine and serotonin (Gesto et al. 2006, 2009). These neurotransmitters are instrumental for proper homeostatic regulation of a wide range of physiological processes including reproduction (Rahman et al. 2011). PAH mechanisms of action on invertebrate reproduction are much less well characterized. However, a few studies on invertebrates have shown effects of PAHs on reproductive output and other whole organism markers used as indices for altered organism development including hatching success rate, recruitment, and time to first reproductive event (Bellas and Thor 2007; Eom et al. 2007; Feldmannová et al. 2006).

PAHs are also known to induce oxidative stress and oxidative stress responses in both invertebrates and vertebrate species following exposure (Dalton et al. 2002; Han et al. 2014; Sun et al. 2006; Valavanidis et al. 2006). Reactive oxygen species (ROS) are produced endogenously in cells, primarily through oxidative metabolism in the mitochondria. However, exposure to contaminants such as PAHs can upset the delicate balance between those processes generating ROS and those responsible for scavenging ROS. Monooxygenase metabolic processes during CYP1a induction following PAH exposure create numerous ROSs including superoxide anions and hydroxyl radicals which are known to deleteriously interact with a host of cellular components (Dalton et al. 2002; Zangar et al. 2004). Some of the more common mechanisms of ROS toxicity are the reaction with, and deterioration of, cell membranes (potentially creating more ROS and mutagenic metabolites), inhibitory impacts on proteins and enzymes, and damage of DNA (Dalton et al. 2002; Kappus 1987; Köhler et al. 2002; Zangar et al. 2004). Compounding the oxidative stress problems associated with the presence of PAHs in aquatic settings is the photoactivation of PAH molecules in the presence of UV, a process which increases the reactivity and toxic potential of some PAHs (Lampi et al. 2006; Spehar et al. 1999; Yu 2003).

A recent development in the understanding of PAH toxicity is the identification of cardiac defects in exposed fish. PAHs are associated with numerous morphological deformities in developing larval fish – including pericardial edema (Barron et al. 2004; Incardona et al. 2011, 2013, 2014). Further analyses have found that cardiac beat frequency, stroke volume, and stroke power are all affected in crude oil-exposed fish (Incardona et al. 2009; Khursigara et al. 2017; Nelson et al. 2016). It has been proposed that PAH exposure during embryonic stages does not cause morphological deformities per se, but rather the changes in cardiac physiology precede the noted morphological defects (Incardona et al. 2004). For example, following exposure to crude oil and to differing PAHs, Brette et al. (2014) determined that certain ionic currents vital for regulating proper myocyte action potential propagation and progression are altered. More specifically, these authors identified PAH-induced blockade of the K^+ currents responsible for cardiomyocyte repolarization and disruption of Ca^+ cycling in the sarcolemma and sarcoplasm reticulum (Brette et al. 2014). These effects on cardiomyocytes change the contractile properties of myocyte sarcomeres, causing the changes in cardiac function observed in organisms following PAH exposure. Similarly, three-ringed phenanthrene has been shown to interact with specific voltage-gated ion channels instrumental for generating ionic currents required for cardiomyocyte action potentials (Brette et al. 2017), lending support to this hypothesis.

3.3.2 Volatiles

Despite their high aqueous solubility and moderate Henry's law constant (a proportionality constant describing the ability of a gas to be dissolved within a liquid given the partial pressure of the gas in the liquid; Mackay and Shiu 1981), volatile organic compounds (VOCs) have relatively low boiling points and high vapor pressures and therefore do not persist in the aquatic environment (Borden et al. 2002; Short 2003). During crude oil releases, there is a systematic and rapid loss of smaller organic compounds, such as benzene, toluene, ethylbenzene, and xylene (collectively termed BTEX), along with their alkyl homologues and other smaller ringed PAHs as weathering increases (Mackay and McAuliffe 1988; Ròques et al. 1994). However, volatile chemicals have been shown to elicit potent acute toxicological effects in aquatic organisms (Mayer and Reichenberg 2006; Neff et al. 2000) and deserve attention as potential chemicals of concern worth reviewing in the context of immediate hazards to organisms associated with surface water FPW contamination events. In this review, we identified the VOCs found in FPW to include differing chlorocarbons, benzene (and related benzene compounds), carbon tetrachloride, styrene, toluene, and xylene (Table 4). Of the above stated VOCs, most are above CEQ environmental water quality guidelines for the protection of aquatic life (Table 4). However, only toluene has been recorded at average concentrations in FPW above LC_{50} values for larval rainbow trout (Table 8). However, benzene concentrations have been found at or near rainbow trout LC_{50} values, while upper-range concentrations of toluene and xylenes (total) occur at or surpass LC_{50} values for almost all aquatic organisms analyzed in this review (Table 8). Despite the highly variable nature of FPW, concentrations found in FPW suggest VOCs are more likely to toxicologically impact aquatic organisms over acute exposure periods (pending on the turbidity and flow dynamic s of the water systems FPW may be released into).

When exposed to water-accommodated fractions of gasoline containing high levels of BTEX, tested species of fish have displayed CYP1a induction and oxidative stress (Schein et al. 2008; Simonato et al. 2011; Stagg et al. 2000). Oxidative stress in these studies was measured in both liver and gill tissue, indicating that multiple organ systems in exposed fish are compromised following exposure to BTEX chemicals (Simonato et al. 2011). Monocyclic aromatic hydrocarbons, like BTEX, are thought to elicit most of their toxicity through induced nonpolar narcosis, a pathology characterized by the insertion of these chemicals into the membranes of cells and subsequent disruption of membrane homeostasis and integrity (Peterson 1994; Ren 2002). As a result, VOC exposures elicit many different physiological

Table 8 Reported acute lethal toxicities (LC_{50} in mg/L) of volatile and synthetic organic chemicals associated with hydraulic fracturing flowback and produced water to aquatic species

Chemical	Species	LC_{50}		Water temp. (°C)	Hardness (mg/L $CaCO_3$)	pH	Reference
		48 h	96 h				
1,1,1-Trichloroethane	DM	>530[a]	–	22	173	8	LeBlanc (1980)[b]
1,1,2-Trichloroethane	DM	11.0[a]	–	22	~115	7.8	U.S. EPA (1978)[c]
		18.0[a]	–	22	173	8	LeBlanc (1980)[b]
		43.0[a]	–	20	Hard	8	Adema and Vink (1981)[c]
		186.0[a]	–	20	~54	~7.2	U.S. EPA (1984)[c]
	Juv. FHM	81.6[d]	81.6[d]	25	45	7.4	U.S. EPA (1984)[c]
		–	81.7[d]	25	56	7.5	Veith et al. (1983)[c]
1,2-Dichloroethane	DM	220.0[a]	–	22	173	8	LeBlanc (1980)[b]
		268.0[a]	–	20	~54	~7.2	U.S. EPA (1984)[c]
	Juv. FHM	118.0[d]	116.0[d]	25	45	7.4	U.S. EPA (1984)[c]
		–	118.0[b]	25	56	7.5	Veith et al. (1983)[c]
	Larval RT	–	~34.0[d]	13	104	8	Black et al. (1982)[c]
1,2-Dichloropropane	Juv. DM	52.0[a]	–	22	173	8	LeBlanc (1980)[b]
	DM	52.5[a]	–	22	~115	7.8	U.S. EPA (1978)[c]
	Juv. FHM	154.0[d]	140.0[d]	25	45	7.4	U.S. EAP (1984)[c]
1,2,4-Trichlorobenzene	DM	2.09[a]	–	20	~54	~7.4	U.S. EPA (1984)[c]
		50.0[a]	–	22	173	8	LeBlanc (1980)[b]
	Juv. FHM	–	2.76[d]	25	45	7.4	U.S. EPA (1984)[c]
		–	2.9[d]	25	56	7.5	Veith et al. (1983)[c]
	Juv. RT	–	1.52[d]	12	~54	~7.4	U.S. EPA (1984)[c]

Benzene	Juv. DM	400.0[a]	–	19	–	7.9	Slooff et al. (1983)[b]
		426.0[a]	–	20	–	7.9	Canton and Adema (1978)[b]
	DM	200.0[a]	–	22	173	8	LeBlanc (1980)[b]
		682.0[a]	–	19	134	7.6	Eastmond et al. (1984)[b]
	DP	–	15.0[a]	15	–	7.5	Trucco et al. (1983)[c]
	Juv. FHM	32.0[a]	32.0[a]	25	360	8.2	Pickering and Henderson (1966)[b]
		35.1[a]	33.5[a]	25	20	7.5	Pickering and Henderson (1966)[b]
		84.0[a]	–	20	140	–	Slooff et al. (1983)[b]
	FHM	–	~15.1[d]	15	–	–	DeGraeve et al. (1982)[c]
	Larval RT	–	8.25[d]	13	94	7.8	Black et al. (1982)[c]
	Juv. RT	–	5.3[d]	15	–	–	DeGraeve et al. (1982)[c]
		56.0[a]	–	15	285	7.5	Slooff et al. (1983)[b]
Carbon tetrachloride	DM	10–100[a]	–	22	~115	7.8	U.S. EPA (1978)[b]
		35.0[a]	–	22	173	8	LeBlanc (1980)[b]
	Larval FHM	–	4.0[d]	20	96	7.8	Black et al. (1982)[c]
	Larval RT	–	1.97[d]	13	104	7.9	Black et al. (1982)[c]
Ethylbenzene	DM	75.0[a]	–	22	173	8	LeBlanc (1980)[b]
	Juv. FHM	42.3[a]	42.3[a]	25	360	8.2	Pickering and Henderson (1966)[b]
		48.5[a]	48.5[a]	25	20	7.5	Pickering and Henderson (1966)[b]
Styrene	DM	23.0[a]	–	22	173	8	LeBlanc (1980)[b]
	Juv. FHM	53.6[a]	46.4[a]	25	20	7.5	Pickering and Henderson (1966)[b]
		62.8[a]	59.3[a]	25	360	8.2	Pickering and Henderson (1966)[b]

(continued)

Table 8 (continued)

Chemical	Species	LC_{50}		Water temp. (°C)	Hardness (mg/L $CaCO_3$)	pH	Reference
		48 h	96 h				
Toluene	Juv. DM	19.6[a,e]	–	20	26	~7.9	Pearson et al. (1979)[b]
	DM	310.0[a]	–	22	173	8	LeBlanc (1980)[b]
	Juv. FHM	–	12.6[d]	20	26	~7.9	Pearson et al. (1979)[b]
		46.3[a]	34.3[a]	25	20	7.5	Pickering and Henderson (1966)[b]
		56.0[a]	42.3[a]	25	360	8.2	Pickering and Henderson (1966)[b]
	Larval RT	–	0.02[d]	14	106	7.8	Black et al. (1982)[c]
	Juv. RT	–	24.0	12	–	–	Johnson and Finley (1980)[b]
Xylene(s)	Juv. FHM	27.7[a]	27.7[a]	25	20	7.5	Pickering and Henderson (1966)[b]
		28.8[a]	28.8[a]	25	360	8.2	Pickering and Henderson (1966)[b]
	Larval RT	–	3.77[d]	13	96	7.8	Black et al. (1982)[b]
	Juv. RT	–	8.2	12	–	–	Johnson and Finley (1980)[b]
Dibromochloromethane	Juv. DM	58.1[a]	46.8[a]	25	Hard	~7.8	Fisher et al. (2014)[c]
Hexachlorobenzene	FHM	–	22.0	20	–	–	Johnson and Finley (1980)[b]
	RT	<0.03[a]	–	15	320	7.4	Calamari et al. (1983)[c]
	ZF	<0.03[a]	–	23	320	7.4	Calamari et al. (1983)[c]

Pentachlorophenol	Juv. DM	0.28[a]	–	19	100	–	Hermens et al. (1984)[c]
		–	0.32[a]	20	130	7.5	Ewell et al. (1986)[b]
		0.48[a]	–	20	–	7.9	Canton and Adema (1978)[b]
		0.48[a]	–	19	–	7.9	Slooff et al. (1983)[b]
		0.6[a]	–	20	–	7.9	Adema (1978)[b]
		1.05[a]	0.8[a]	20	Hard	8	Adema and Vink (1981)[c]
	DM	0.68[a]	–	22	173	8	LeBlanc (1980)[b]
		1.4[a]	0.68[a]	20	Hard	8	Adema and Vink (1981)[c]
	Juv. FHM	0.21[a]	–	20	140	–	Slooff et al. (1983)[b]
		–	0.32[a]	20	130	7.5	Ewell et al. (1986)[b]
		7.9[d]	0.23[d]	25	~46	~7.9	Phipps et al. (1981)[c]
	FHM	–	0.25[a]	22	~170	–	Dwyer et al. (2005)[b]
	Juv. RT	0.2[a]	–	15	285	7.5	Slooff et al. (1983)[b]
		0.093	–	11	45	7.6	McKim et al. (1987)[c]

DM Daphnia magna, DP Daphnia pulex, FHM fathead minnow (*Pimephales promelas*), *RT* rainbow trout (*Oncorhynchus mykiss*), *ZF* zebrafish (*Brachydanio rerio*), *Juv*. juvenile-aged organisms used. DM = within 24 h. RT = ≤ 1-year post hatch

[a]Static/semi-static exposures

[b]Nominal chemical concentrations used for LC_{50} calculations

[c]Measured chemical concentrations used for LC_{50} calculations

[d]Flow-through exposures

[e]EC_{50} Measurement

forms of toxicity in multiple cell, tissue, and organ systems. For example, in spermatozoa exposed to toluene, toxicity has been measured via the production of 7,8-dihydro-20-deoxyguanosine (8-oxodG), a marker of oxidative DNA damage (Nakai et al. 2003), while benzene has been shown to be a potent inducer of ROS including the superoxide radical anions, hydroperoxyl radicals, hydrogen peroxide (H_2O_2), and the highly reactive hydroxyl radicals (Ho and Witz 1997; Kolachana et al. 1993; Wiemels and Smith 1999; Winn 2003). In nonaquatic species studies, these radicals then affect numerous other biological systems including, but not limited to, embryonic development (reviewed in Badham et al. 2010). Interestingly, in fish, BTEX components of crude oil have not been shown to have major discernible effects on embryonic development (Anderson et al. 2009; Jung et al. 2013; Marty et al. 1997). VOCs associated with FPW are also known to be both induced and metabolized by multiple different CYP monooxygenase systems, specifically the CYP2 family and to a lesser extent CYP1A (Nakajima 1997; Stagg et al. 2000). Again, these enzymes may generate reactive oxygen species leading to increased oxidative stress (Dalton et al. 2002; Zangar et al. 2004).

BTEX effects on endocrine-related activities in aquatic species are relatively less well studied. In marine crabs, benzene delays molting, an effect likely mediated through endocrine disruption (Zou 2005). In experimental and epidemiological studies of nonaquatic species, differing BTEX compound exposures have been reported to both stimulate hypothalamic pituitary adrenocortical activity and adrenocorticotropin release and decrease pituitary gonadotropin activity. These findings highlight the diverse nature of BTEX effects on the hypothalamic-pituitary system and suggest that further investigation to elucidate mechanistic modes of action is required (Baccarelli et al. 2000; Verma and Rana 2009).

In humans, male workers exposed to BTEX compounds exhibit reduced semen quality and vitality (Xiao et al. 2001). Repeated exposure to ethylbenzene and benzene has also been shown to induce cancerous tumors and lesions in the testes, respectively, while both toxicant exposures were associated with decreased sperm counts, lowered sperm mobility, and increased sperm DNA damage in male workers (Chan et al. 1998; Katukam et al. 2012; Ward et al. 1985). Toluene has also been shown to induce spermatozoa DNA damage and to play a role in reproductive toxicity (Nakai et al. 2003). In female human and rodent species, BTEX compounds are theorized to affect luteal functions of the ovaries, while benzene has been specifically identified as causing ovarian toxicity and carcinogenicity (Maronpot 1987; Reutman et al. 2002). Therefore, despite the paucity of studies on the effects of VOCs on aquatic organism endocrine systems and reproductivity, the effects described above are likely to also manifest in aquatic organisms should concentrations reach effective concentrations. However, given the volatility of BTEX compounds, it is expected that for many of these endocrine disrupting and reproductive effects (among other toxicological end points) to manifest in aquatic organisms following FPW release, repeated or extended exposures would need to occur.

3.3.3 Synthetics

Our analysis of FPW chemical composition literature has identified that organochlorines are the most characteristic and consistently present synthetic organic compounds in FPW. These compounds include the trihalomethane compound dibromochloromethane (DBCM) and the chlorobenzene compounds hexachlorobenzene (HCB) and pentachlorophenol (PCP). HCB and PCP are typically employed as pesticides, whereas DBCM is a disinfection by-product from the reaction of chlorine with organic matter and bromide ions but has also been used as a solvent and flame retardant (Bailey 2001; Cirelli 1978; Munson et al. 1982). Although we have classified DBCM as "synthetic," it should be noted that DBCM may be produced in small quantities by certain species of marine algae (Itoh and Shinya 1994), and thus, natural production of the compound may occur. As their anthropogenic origin suggests, these synthetic compounds found in FPW are most probably components of the original fracturing fluids injected into the well during the production stage of drilling practices. Of these three synthetic species (HCB, PCP, and DBCM), average concentrations are below concentrations expected to induce significant lethality to aquatic biota (Tables 4 and 8). However, HCB upper-range FPW concentrations are found to correspond to LC_{50} values for rainbow trout and zebrafish, while PCP upper-range FPW concentrations are found at or above LC_{50} values for most aquatic model species examined (Table 8). While limited CEQ and USEPA water quality guideline data exists for DCBM and HCB, PCP concentrations in FPW were found to be above guideline concentrations (Table 4). The toxicity of HCB and PCP are likely driven to a significant extent by their relatively high octanol-water partitioning coefficients (KOWs) and corresponding potential for bioaccumulation (Sabljić et al. 1995; Chamberlain et al. 1996). Thus, a synthetic organic compound toxicological hazard to organisms will likely exist if spills or releases of FPW to surface waters occur.

HCB and PCP toxicity to aquatic organisms has been studied for many years (Goodnight 1942; Johnson et al. 1974). However, only a few sublethal studies pertaining to endocrine system function and reproduction in aquatic species have been performed. In the Chinese rare minnow (*Gobiocypris rarus*), PCP exposure invokes upregulation of certain EDC-related genes (e.g., estrogen receptor alpha and beta, androgen receptor, and vitellogenin) in male hepatic and gonadal tissues. PCP also affects the hypothalamic-pituitary-gonadal/interrenal (HPG/I) axis, resulting in changes to plasma steroid hormone levels and spermatogenesis rates. In female rare minnows, PCP negatively impacted the HPG/I axis and increased estradiol and testosterone plasma concentrations while decreasing cortisol levels. Correspondingly, degenerate ovaries were also characterized (Yang et al. 2017). Changes in vitellogenin in PCP-exposed male Japanese medaka (*Oryzias latipes*) have also been observed, in association with decreased fecundity, fertility, and gonadal health in both male and female fish (Zha et al. 2006). Similarly, serum testosterone levels significantly increased in crucian carp (*Carassius carassius*) following sublethal PCP exposure over 7 and 15 days (Zhang et al. 2008), and detrimental effects to gonad growth and development have been observed in multiple other freshwater fish species (Hanson et al. 2007). In vitro studies have also shown that PCP exhibits estrogenic, antiestrogenic, antiandrogenic, and antagonistic retinoid X receptor effects in a variety of tissue and cell types (Jung et al. 2004; Li et al. 2008; Orton et al. 2009; Suzuki et al. 2001).

In nonaquatic species, HCB has been found to negatively affect steroidogenesis in monkey ovarian tissue, possibly due to peroxidation of the mitochondrial lipid membrane (Foster et al. 1995). In porcine ovarian follicle tissue cultures, HCB inhibits both testosterone and estrogen secretion and reduces expression of key steroidogenic enzymes, such as CYP17, CYP17β-HSD, and CYP19 (Gregoraszczuk et al. 2011). Furthermore, an interplay between HCB and the estrogen receptor α (ERα) has been reported. Following exposure in healthy rats, HCB increases ERα protein expression and 17-β estradiol and prolactin levels and reduces progesterone, follicle-stimulating hormone, and luteotrophic hormone concentrations in mammary tissues (Peña et al. 2012). Interestingly, in vitro studies have also shown that breast cell proliferation and subsequent tumor formation following HCB exposure are ERα dependent, further exemplifying the interplay between HCB and the ERα (García et al. 2010).

Similar to VOCs, nonpolar narcosis is one theorized mode of toxicity associated with HCB and PCP exposure in aquatic organisms (Call et al. 1985; Ren 2002). These narcotic effects may result in a myriad of impacts to biological systems due to their deleterious effects on cell lipid membranes and lipid membrane functions. PCP is also a known oxidative phosphorylation uncoupler and causes mitochondrial membrane damage (Weinbach 1954; Weinbach and Garbus 1969). Accordingly, PCP (and its metabolite tetrachlorohydroquinone) has been shown to induce ROS in rainbow trout hepatocyte cell lines while also causing lipid peroxidation, DNA damage, and ROS induction in zebrafish larvae (*Danio rerio*) (Fang et al. 2015; Pietsch et al. 2014). Similarly, HCB has been shown to cause oxidative stress in common carp (*Cyprinus carpio*), where brain tissue in exposed organisms displayed elevated levels of ROS, thiobarbituric acid reactive substances (indicative of lipid peroxidation), and nitric oxide synthase (NOS) activity (Song et al. 2006). While PCP has not been conclusively shown to bind the AhR, HCB has been shown to weakly associate with the AhR, and both PCP and HCB are shown to induce CYP1A and ethoxyresorufin-O-deethylase (a marker for CYP1A induction) activity in a variety of organisms and tissue/cell models (Mundy et al. 2010, 2012; Starek-Świechowicz et al. 2017; Zha et al. 2006; Zhang et al. 2008). Again, induction of these monooxygenase detoxification pathways may contribute to the overall increased oxidative stress observed in organisms exposed to these two chemicals.

3.4 Whole FPW Toxicity

To date, very few studies have examined the toxicity of raw FPW samples. Some groups have taken characterized components of FPW and reconstituted them into mixtures in an attempt to simulate FPW from a representative well (Boulé et al. 2018; Kassotis et al. 2018b). Other studies have sampled for water affected by UOG operations and tested these waters for adverse effects on organisms (Tasker et al. 2018). However, the highly complex nature of FPW fluids may render studies examining reconstituted solutions or waters affected by UOG activities less practical for accurately estimating full toxicological effects and determining mechanistic modes of toxicity. Under laboratory settings, whole effluent toxicity (WET) studies are better able to control for environmental variables, determine accurate dose effects, and observe real-time events leading to toxicity in exposed organisms (Chapman 2000). Examples of where such WET studies have significantly contributed to the understanding of specific detrimental environmental incidents are studies pertaining to oil spills, such as the Exxon Valdez and the Deepwater Horizon releases (Alloy et al. 2016; Heintz et al. 1999). Because of such WET studies, our understanding of crude oil impacts on multiple levels of biological organization (cell, tissue, organs, and whole organism) has significantly advanced, and risk assessment and adverse outcome pathway analyses now exist should future spills of crude oil occur (Ankley et al. 2010; Incardona and Scholz 2016; Short 2003). Thus, investigation on whole, raw FPW samples is required to gain a better understanding of the toxicological impacts to the biota when releases of FPW to the environment occur.

Studies which utilize either simulated or field-collected UOG-affected samples may, however, offer insight into possible pathways of toxicity which are induced when FPW exposure to organisms occurs. In a study using groundwater samples collected from regions experiencing horizontal hydraulic fracturing activities in Wyoming, USA, greater ER antagonist activity and increased progesterone receptor antagonist activity were exhibited using a human endometrial cell gene reporter assay (Kassotis et al. 2018b). These groundwater samples contained organic contaminants, such as 2-ethylhexanol and styrene, which were suggested to be responsible for the bioactivity observed. Supporting the notion that development and endocrine activity may be impacted following exposure to FPW, a review performed by Webb et al. (2014) identified other compounds in FPW which may cause sublethal developmental and endocrine disruption effects.

Other endpoints associated with exposure to FPW include markers of xenobiotic metabolism and an altered immune system. Using surface samples of FPW collected following spreading on roads for de-icing or dust suppression purposes, increased bioassay activity for both AhR and pregnane X receptor (PXR) activity was observed in both daphnid and in reporter cell lines (Tasker et al. 2018). Similarly, using a mixture of 23 chemicals associated with UOG operations, developmental exposures in mice were found to significantly alter sex-specific immunological responses (e.g., T cell-dependent allergic airway disease responses and influenza A virus infection leukocyte recruitment) and intensify autoimmune encephalitis responses in exposures as low as 0.1 and 1.0 μg/mL dosages (Boulé et al. 2018). Furthermore, some of these effects were observed to persist weeks postexposure. A similar simulated FPW mixture, also consisting of 23 distinct FPW-related chemicals, was used to expose *Xenopus laevis* tadpoles which had contracted the ranavirus FV3. Mixture exposures of <5.0 μg/L significantly affected tadpole homeostatic myeloid lineage gene expression and increased viral load (Robert et al. 2018). Changes in the key immunological genes, TNF-α, IL-1β, and Type I IFN, were also observed. These findings suggest that FPW-associated contaminants may acutely disrupt immune function in organisms exposed at low doses.

Another pathway demonstrated to be affected following exposure to simulated FPW is metabolic function. Adipose tissue function in mice was examined using a 23-compound simulated mixture of chemicals associated with UOG operations and a small subset of surface water samples contaminated with UOG wastewater from Colorado and West Virginia (Kassotis et al. 2018a). These researchers showed that adipogenic activity (both triglyceride accumulation and pre-adipocyte proliferation) in a mouse adipose cell line was potently induced following exposure. Furthermore, exposure to contaminated surface water samples resulted in increased peroxisome proliferator-activated receptor gamma activity (a receptor critical for regulating fatty acid storage and glucose metabolism) in a human embryonal kidney cell line reporter assay (Kassotis et al. 2018a). Unfortunately, studies investigating the effects of FPW on metabolism in aquatic animals are lacking.

Regarding whole, actual FPW toxicological analyses performed on freshwater species, it is suggested that the high saline content typical of many FPW samples elicits most of the associated toxicity in exposed organisms (Blewett et al. 2017c; Folkerts et al. 2019; He et al. 2017a). Despite this saline-induced toxicity, trends in lethality are highly species dependent and may be subject to the content of other contaminants present in FPW, including the specific organic, metal, and other chemical species profiles which change depending on the geology of the formation, well-operation parameters, and time of flowback. For example, in a study performed on FPW samples sequentially sampled over time from a single well, *Daphnia magna* were observed to be most sensitive of the species tested (lowest LC_{50} value of 0.76% dilution of whole FPW), while rainbow trout embryos were found to be least sensitive (lowest LC_{50} value of 12.26% dilution of whole FPW) (Folkerts et al. 2019). This high sensitivity to FPW exposure observed in *Daphnia magna* matches findings of a similar study investigating *Daphnia magna* reproductive alterations following FPW exposure (Blewett et al. 2017a). Interestingly, sampling time-dependent changes in FPW chemistry also mediate significant changes in toxicity in most exposed species of organisms (Folkerts et al. 2019), implying that other components besides salt (e.g., organics, other anthropogenic contaminants) add to the overall toxicological profile of FPW and should be considered for future risk assessment and hazard identification policy development.

Recent studies have also identified mechanisms of FPW toxicity upon exposure of aquatic animals to raw FPW samples collected from well heads. One of the first studies to use raw FPW for WET analyses showed that sublethal concentrations of raw FPW and sediment-free fractions of FPW affected xenobiotic metabolism, induced oxidative stress, and caused endocrine disrupting effects in rainbow trout gill and liver tissue (He et al. 2017b). Further work on rainbow trout reaffirmed the effects of FPW on oxidative stress pathways but also showcased how gill morphology is negatively impacted following FPW exposure. Chief among these gill morphological alterations were observations of increased interlamellar cell mass measurements, an effect resulting from the highly saline nature of FPW solutions (Blewett et al. 2017c). Corroborating these rainbow trout-induced oxidative stress observations, specific reporter cell line assays have additionally shown marked oxidative (and genotoxic) responses following FPW exposure (Faber et al. 2019). Observations of xenobiotic metabolism induction were also observed in zebrafish embryos/larvae exposed to sublethal concentrations of FPW (He et al. 2017a, 2018). These studies highlight the negative impacts of this fluid on zebrafish early development, evidenced by increased occurrences of larvae developmental deformities such as pericardial edema and tail/spine curvature. Similar embryonic studies on zebrafish exposed to FPW have also implicated the cardio-respiratory system as a biological endpoint affected following FPW exposure (Folkerts et al. 2017a, b). When exposed embryonically (at 24 h post fertilization; hpf) to sublethal concentrations of raw FPW and created sediment-free FPW fractions, zebrafish embryos display reduced O_2 consumption and altered expression of genes associated with cardiac function and development (Folkerts et al. 2017b). Following postembryonic exposure, larval staged fish additionally showcased higher rates of morphological deformities including pericardial edema, yolk sac edema, and tail/spine curvature and had reduced heart rates, relative to controls. Using a similar exposure regimen of exposing embryos at 24 hpf to raw FPW and sediment-free fractions, additional latent effects on the cardio-respiratory in juvenile zebrafish have been observed. After 24 and 48 h embryonic exposures to 2.5 and 5% raw FPW (and sediment-free FPW fractions), juvenile zebrafish 60 days postexposure displayed reduced critical swim speeds (a measure of swim performance) and had reduced aerobic capacities (Folkerts et al. 2017a).

Sublethal effects of raw, real-world FPW samples on species other than fish have also been recorded. It has been observed that with some FPW samples, the small freshwater crustacean *Daphnia magna* can become trapped at the surface of FPW contaminated waters at low dilutions of 0.01% (Blewett et al. 2018). This was proposed, although not demonstrated, to be a consequence of surfactants in the FPW sample, reducing water surface tension. As *Daphnia* broke through the surface, they become trapped in oil deposits and FPW-related detritus and could not return to the water column, resulting in physical immobility (Blewett et al. 2018). In studies where *Daphnia* remained in the water column, raw FPW was observed to induce biotransformative processes, reduce oxygen consumption, and negatively impact reproductive output at low dilutions as low as 0.008 and 0.04% (Blewett et al. 2017a).

4 FPW Releases and Regulatory Guideline Considerations

The amount of FPW produced by hydraulic fracturing activities in North America is substantial and growing. For example, it is estimated that 210 billion gallons of FPW were produced from 2005 to 2014 in the USA alone (Kondash and Vengosh 2015). In Canada, formations exploited for shale gas and tight oil extraction using UOG technologies primarily occur in the western provinces of British Columbia (BC) and Alberta (AB). These formations include, among others, the Montney and Duvernay. In the Montney formation, approximately 50–100% of initial fracturing fluid/water injected is recovered as FPW (Rivard et al. 2014), and between 2011 and 2013, it was estimated that 10,000–25,000 m^3 of wastewater was produced per well in this formation (Goss et al. 2015). The volumes of FPW produced from the Duvernay formation are less well characterized, but estimates of water use in 2013 found volumes of 10,000–60,000 m^3 of water per well (Goss et al. 2015). FPW generation and management are growing concerns among regulatory bodies in regions utilizing UOG practices, and aspects of UOG development pertaining to FPW characterization and spill hazard assessment are only recently being analyzed and studied more intimately (reviewed in Alessi et al. 2017; Goss et al. 2015). Considering UOG activities are anticipated to increase in prevalence in North America (Gagnon et al. 2016; Vengosh et al. 2014), determining how to effectively gauge the severity of spills and the extent of impact to the environment is a topic of increasing importance.

In terms of spills and their impacts, surface water contamination of FPW is of main concern, although groundwater and aquifer contamination events are of similar concern to certain regions experiencing heavy UOG activity. Common routes of FPW contamination to surface water sources include accidental releases during FPW transport (i.e., pipeline or truck transport), inadequate FPW treatment and subsequent release, and intentional illegal releases by operators (Goss et al. 2015; Vengosh et al. 2014). From 2005 to 2010, it was determined that within the Marcellus and Fayetteville shale plays of the eastern USA, the average distance of drilled wells from freshwater streams was ~300 m (Entrekin et al. 2011). Given the close association of hydraulic fracturing activities with freshwaters and the growth of this extraction approach, the potential for releases to surface bodies of water is expected to rise. In AB alone, there were a reported 205 spills/releases of salt/produced water associated with UOG or crude oil production from January 2017 to July 2018 (data collected from the AER incidents compliance dashboard). The BC Oil and Gas Commission (BCOGC) reported that nine spills of emulsion (oil, gas, water, and condensate) and three spills of brine from UOG activities occurred during 2017 in and around the Montney shale formation of upper northeast BC (data collected from the BCOGC n.d.). Furthermore, in the state of Colorado, there were a reported 257 spills of produced water in 2016 (data from the Center for Western Priorities 2016). To exemplify how spills are of concern and FPW management strategies need to be more carefully considered among regulatory agencies and operators, a report was released by the United States Environmental Protection Agency in 2015 studying hydraulic fracturing spills from January 2006 to

April 2012 in the states of Colorado, Pennsylvania, Oklahoma, Arkansas, Texas, Louisiana, New Mexico, Wyoming, Utah, West Virginia, and North Dakota. This report found a total of combined 24,000 hydraulic fracturing-related reported spills at sites located at a distance from well pads and an additional 457 spills at well pad sites (U.S. EPA 2016).

One problem that immediately limits spill remediation efforts and hinders research of FPW effects on the ecosystem is that neither Canada or the USA classifies wastewater produced from hydraulic fracturing as hazardous waste (Goss et al. 2015). This is an issue that impedes our ability to properly estimate and attempt to mitigate the destructive impacts to the environment when releases of FPW occur and which is likely to limit investment into these efforts on behalf of the industry. Combine this with the largely decentralized nature of governance over hydraulic fracturing practices (most hydraulic activity is governed by state by state or provincial natures), a highly variable system of policies and protocols devoted to spill and remediation regulation results (Goss et al. 2015; Ralston and Kalmbach 2018). Accordingly, requirements for spill mitigation are not consistently included among regional regulatory guidelines and are not currently observed to be greatest drivers of risk (Kulander 2013; Richardson et al. 2013). Thus, while it is acknowledged that practices and regulations have improved over the past decade, and that the UOG industry allocates substantial resources toward for both spill prevention and for mitigation efforts when releases of FPW to surface bodies of water occur, it is clear that federally mandated remediation guidelines are required. A comprehensive analysis of all region's management practices, along with policy shortfalls, would help develop a cohesive set of guidelines that address proper remediation strategies and ensure that all industry players, regardless of region, comply to a set of standard mitigation measures that ensures protection/remediation of the aquatic environment.

5 Conclusions and Future Directions

Technological advancements in the oil and gas sectors have led to companies being able to exploit formations that were previously thought to be of limited potential for hydrocarbon extraction. The rapid rise of hydraulic fracturing methodologies, including UOG methodologies such as horizontal hydraulic fracturing, for oil and gas extraction, has resulted in an expansion of reserve development. However, it has also increased concern about risks to the environment. These concerns include water sourcing availability, increased seismic activity hazards, and groundwater contamination. In the current review, we have also addressed the environmental concern of FPW pollution to surface water and the potential adverse effects on animals inhabiting these freshwater bodies. FPW has the potential to be the next great industrial wastewater management conundrum given the extensive volume produced per well, the vast amount of producible reserves in North America, and the expectation of continued UOG expansion. After examining thousands of records of FPW characterizations from wells located in predominantly developed North American shale plays, we provide a review that explores both lethal and sublethal responses to FPW in commonly studied toxicological model aquatic species. We found that certain chemical species may be found in concentrations high enough to result in lethal or significant sublethal responses in organisms which encounter FPW. While this review focused on effects at an organismal level, it is important to note that changes at the level of an individual can have long-term impacts on populations and ecosystems, which may extend effects beyond the specific period of the initial FPW contamination event. Elements of this review may serve to identify certain key points required for effective FPW risk assessment development and improve release identification, assessment, and remediation protocols.

Acknowledgments This work is funded by the Natural Sciences and Engineering Research Council (NSERC) Collaborative Research and Development Grant CRDPJ 460308 – 14 and co-funded by Encana Services Company Ltd. to Dr. Greg Goss. We would like to thank the countless number of students and staff who have worked on various FPW projects over the years which have significantly helped advance our understanding and garner interest in the field of FPW toxicity. We, the authors, declare no conflict of interest.

References

Adema D (1978) Daphnia magna as a test animal in acute and chronic toxicity tests. Hydrobiologia 59:125. https://doi.org/10.1007/bf00020773

Adema D, Vink I (1981) A comparative study of the toxicity of 1, 1, 2-trichloroethane, dieldrin, pentachlorophenol and 3, 4 dichloroaniline for marine and freshwater organisms. Chemosphere 10:533. https://doi.org/10.1016/0045-6535(81)90255-1

Aiken GR, Hsu-Kim H, Ryan JN (2011) Influence of dissolved organic matter on the environmental fate of metals, nanoparticles, and colloids. Environ Sci Technol 45:3196–3201. https://doi.org/10.1021/es103992s

Alberta Energy Regulator (n.d.) Compliance dashboard (AER). Suite 1000, 250–5 Street SW, Calgary, Alberta, T2P 0R4. http://www1.aer.ca/compliancedashboard/incidents.html. Accessed Aug 2018

Alessi DS, Zolfaghari A, Kletke S, Gehman J (2017) Comparative analysis of hydraulic fracturing wastewater practices in unconventional shale development: water sourcing, treatment and disposal practices. Can Water Resour J 42:105–121. https://doi.org/10.1002/9781119099987.ch4

Alexander HC, Quick JA, Bartlett EA (1981) Static acute toxicity of sodium bromide to fathead minnows. Bull Environ Contam Toxicol 27:326. https://doi.org/10.1007/bf01611027

Allen EW (2008) Process water treatment in Canada's oil sands industry: I. Target pollutants and treatment objectives. J Environ Eng Sci 7:123–138. https://doi.org/10.1139/s07-038

Alloy M, Baxter D, Stieglitz J, Mager E, Hoenig R, Benetti D, Grosell M, Oris J, Roberts A (2016) Ultraviolet radiation enhances the toxicity of deepwater horizon oil to Mahi-mahi (Coryphaena hippurus) embryos. Environ Sci Technol 50:2011–2017. https://doi.org/10.1021/acs.est.5b05356

Alsop DH, Wood CM (1999) Influence of waterborne cations on zinc uptake and toxicity in rainbow trout, Oncorhynchus mykiss. Can J Fish Aquat Sci 56:2112–2119

Alsop DH, McGeer JC, McDonald DG (1999) Costs of chronic waterborne zinc exposure and the consequences of zinc acclimation on the gill/zinc interactions of rainbow trout in hard and soft water. Environ Toxicol Chem 18:1014. https://doi.org/10.1002/etc.5620180529

Anderson BS, Arenella-Parkerson D, Phillips BM, Tjeerdema RS, Crane D (2009) Preliminary investigation of the effects of dispersed Prudhoe Bay Crude Oil on developing topsmelt embryos, Atherinops affinis. Environ Pollut 157:1058–1061. https://doi.org/10.1016/j.envpol.2008.10.013

Ankley GT, Bennett RS, Erickson RJ, Hoff DJ, Hornung MW, Johnson RD, Mount DR, Nichols JW, Russom CL, Schmieder PK, Serrrano JA, Tietge JE, Villeneuve DL (2010) Adverse outcome pathways: a conceptual framework to support ecotoxicology research and risk assessment. Environ Toxicol Chem 29:730–741. https://doi.org/10.1002/etc.34

Annevelink MPJA, Meesters JAJ, Hendriks AJ (2016) Environmental contamination due to shale gas development. Sci Total Environ 550:431–438. https://doi.org/10.1016/j.scitotenv.2016.01.131

Baccarelli A, Pesatori AC, Bertazzi PA (2000) Occupational and environmental agents as endocrine disruptors: experimental and human evidence. J Endocrinol Investig 23:771–781. https://doi.org/10.1007/BF03345069

Badham HJ, Renaud SJ, Wan J, Winn LM (2010) Benzene-initiated oxidative stress: effects on embryonic signaling pathways. Chem Biol Interact 184:218–221. https://doi.org/10.1016/j.cbi.2009.11.005

Bailey RE (2001) Global hexachlorobenzene emissions. Chemosphere 43:167–182

Baird DJ, Barber I, Soares A, Calow P (1991) An early life-stage test with Daphnia magna Straus: an alternative to the 21-day chronic test? Ecotoxicol Environ Saf 22(1):1–7. https://doi.org/10.1016/0147-6513(91)90040-v

Barron MG, Carls MG, Heintz R, Rice SD (2004) Evaluation of fish early life-stage toxicity models of chronic embryonic exposures to complex polycyclic aromatic hydrocarbon mixtures. Toxicol Sci 78:60–67. https://doi.org/10.1093/toxsci/kfh051

Baumann PC, Smith IR, Metcalfe CD (1996) Linkages between chemical contaminants and tumors in Benthic Great Lakes Fish. J Great Lakes Res 22:131–152. https://doi.org/10.1016/s0380-1330(96)70946-2

Bellas J, Thor P (2007) Effects of selected PAHs on reproduction and survival of the calanoid copepod Acartia tonsa. Ecotoxicology 16:465–474. https://doi.org/10.1007/s10646-007-0152-2

Benoit DA, Holcombe GW (1978) Toxic effects of zinc on fathead minnows Pimephales promelas in soft water. J Fish Biol 13:701. https://doi.org/10.1111/j.1095-8649.1978.tb03484.x

Bertilsson S, Widenfalk A (2002) Photochemical degradation of PAHs in freshwaters and their impact on bacterial growth–influence of water chemistry. Hydrobiologia 469:23–32

Biesinger KE, Christensen GM (1972) Effects of various metals on survival, growth, reproduction, and metabolism of Daphnia magna. J Fish Res Board Can 29:1691. https://doi.org/10.1139/f72-269

Bilbao E, Raingeard D, de Cerio OD, Ortiz-Zarragoitia M, Ruiz P, Izagirre U, Orbea A, Marigómez I, Cajaraville MP, Cancio I (2010) Effects of exposure to Prestige-like heavy fuel oil and to perfluorooctane sulfonate on conventional biomarkers and target gene transcription in the thicklip grey mullet Chelon labrosus. Aquat Toxicol 98:282–296. https://doi.org/10.1016/j.aquatox.2010.02.018

Black JA, Birge WJ, McDonnell WE, Westerman AG (1982) Aquatic toxicity of organic compounds to embryo-larval stages of fish and amphibians. University of Kentucky Water Resources Research Institute. Report No. 133. https://doi.org/10.1520/stp34883s

Blair SD, Matheson D, He Y, Goss GG (2016) Reduced salinity tolerance in the Arctic grayling (Thymallus arcticus) is associated with rapid development of a gill interlamellar cell mass: implications of high-saline spills on native freshwater salmonids. Conserv Physiol 4:1–11. https://doi.org/10.1093/conphys/cow010

Blauch ME, Myers RR, Moore TR, Lipinski BA (2009) Marcellus shale post-frac flowback waters – where is all the salt coming from and what are the implications? Society of Petroleum

Engineers International 125740. https://www.onepetro.org/conference-paper/SPE-125740-MS. Accessed Mar 2018

Blewett TA, Leonard EM (2017) Mechanisms of nickel toxicity to fish and invertebrates in marine and estuarine waters. Environ Pollut 223:311–322. https://doi.org/10.1016/j.envpol.2017.01.028

Blewett TA, Delompré PLM, He Y, Folkerts EJ, Flynn SL, Alessi DS, Goss GG (2017a) Sublethal and reproductive effects of acute and chronic exposure to flowback and produced water from hydraulic fracturing on the water Flea Daphnia magna. Environ Sci Technol 51:3032–3039. https://doi.org/10.1021/acs.est.6b05179

Blewett TA, Simon RA, Turko AJ, Wright PA (2017b) Copper alters hypoxia sensitivity and the behavioural emersion response in the amphibious fish Kryptolebias marmoratus. Aquat Toxicol 189:25–30. https://doi.org/10.1016/j.aquatox.2017.05.007

Blewett TA, Weinrauch AM, Delompré PLM, Goss GG (2017c) The effect of hydraulic flowback and produced water on gill morphology, oxidative stress and antioxidant response in rainbow trout (Oncorhynchus mykiss). Sci Rep 7:46582. https://doi.org/10.1038/srep46582

Blewett TA, Delompré PLM, Glover CN, Goss GG (2018) Physical immobility as a sensitive indicator of hydraulic fracturing fluid toxicity towards Daphnia magna. Sci Total Environ 635:639–643. https://doi.org/10.1016/j.scitotenv.2018.04.165

Boeckman CJ, Bidwell JR (2006) The effects of temperature, suspended solids, and organic carbon on copper toxicity to two aquatic invertebrates. Water 171:185. https://doi.org/10.1007/s11270-005-9036-3

Borden RC, Black DC, McBlief KV (2002) MTBE and aromatic hydrocarbons in North Carolina stormwater runoff. Environ Pollut 118:141–152

Boudet H, Clarke C, Bugden D, Maibach E, Roser-Renouf C, Leiserowitz A (2014) "Fracking" controversy and communication: using national survey data to understand public perceptions of hydraulic fracturing. Energy Policy 65:57–67. https://doi.org/10.1016/j.enpol.2013.10.017

Boulé LA, Chapman TJ, Hillman SE, Kassotis CD, O'Dell C, Robert J, Georas SN, Nagel SC, Lawrence BP (2018) Developmental exposure to a mixture of 23 chemicals associated with unconventional oil and gas operations alters the immune system of mice. Toxicol Sci 163:639–654. https://doi.org/10.1093/toxsci/kfy066

Boullemant A, Lavoie M, Fortin C, Campbell PGC (2009) Uptake of hydrophobic metal complexes by three freshwater algae: unexpected influence of pH. Environ Sci Technol 43:3308–3314

Bradley RW, Sprague JB (1985) The influence of pH, water hardness, and alkalinity on the acute lethality of zinc to rainbow trout (Salmo gairdneri). Can J Fish Aquat Sci 42:731–736. https://doi.org/10.1139/f85-094

Brette F, Machado B, Cros C, Incardona JP, Scholz NL, Block BA (2014) Crude oil impairs cardiac excitation-contraction coupling in fish. Science 343:772–776. https://doi.org/10.1126/science.1242747

Brette F, Shiels HA, Galli GLJ, Cros C, Incardona JP, Scholz NL, Block BA (2017) A novel cardiotoxic mechanism for a pervasive global pollutant. Sci Rep 7:41476. https://doi.org/10.1038/srep41476

British Columbia Oil and Gas Commission (BCOGC) (n.d.) BCOGC incident map. https://www.bcogc.ca/public-zone/bcogc-incident-map. Accessed Aug 2018

Brungs WA (1969) Chronic toxicity of zinc to the fathead minnow, Pimephales promelas Rafinesque. Trans Am Fish Soc 98:272–279. https://doi.org/10.1577/1548-8659(1969)98[272:CTOZTT]2.0.CO;2

Buhler DR, Williams DE (1988) The role of biotransformation in the toxicity of chemicals. Aquat Toxicol 11:19–28. https://doi.org/10.1016/0166-445x(88)90004-5

Bury NR, Walker PA, Glover CN (2003) Nutritive metal uptake in teleost fish. J Exp Biol 206:11–23

Büsselberg D (1995) Calcium channels as target sites of heavy metals. Toxicol Lett 82–83:255–261

Cairns MA, Nebeker AV (1982) Toxicity of acenaphthene and isophorone to early life stages of fathead minnows. Arch Environ Contam Toxicol 11:703–707. https://doi.org/10.1007/bf01059158

Calamari D, Marchetti R, Vailati G (1980) Influence of water hardness on cadmium toxicity to *Salmo gairdneri* rich. Water Res 14:1421–1426. https://doi.org/10.1016/0043-1354(80)90006-8

Calamari D, Galassi S, Setti F, Vighi M (1983) Toxicity of selected chlorobenzenes to aquatic organisms. Chemosphere 12:253–262. https://doi.org/10.1016/0045-6535(83)90168-6

Call DJ, Brooke LT, Knuth ML, Poirier SH, Hoglund MD (1985) Fish subchronic toxicity prediction model for industrial organic chemicals that produce narcosis. Environ Toxicol Chem 4:335–343. https://doi.org/10.1897/1552-8618(1985)4[335:fstpmf]2.0.co;2

Canton JH, Adema D (1978) Reproducibility of short-term and reproduction toxicity experiments with Daphnia magna and comparison of the sensitivity of Daphnia magna with Daphnia pulex and Daphnia cucullata in short-term experiments. Hydrobiologia 59:135–140

Canton JH, Slooff W (1982) Toxicity and accumulation studies of cadmium (Cd2+) with freshwater organisms of different trophic levels. Ecotoxicol Environ Saf 6:113–128

Canton JH, Wester PW (1983) Study on the toxicity of sodium bromide to different freshwater organisms. Food Chem Toxicol 21:369–378. https://doi.org/10.1016/0278-6915(83)90090-x

Center for Western Priorities (2016) Colorado oil and gas toxic release tracker. http://westernpriorities.org/2016-colorado-oil-and-gas-toxic-release-tracker/. Accessed Aug 2018

Chamberlain K, Evans AA, Bromilow RH (1996) 1-octanol/water partition coefficient (Kow) and pKa for ionisable pesticides measured by a pH-metric method. Pestic Sci 47:265–271. https://doi.org/10.1002/(sici)1096-9063(199607)47:3

Chan PC, Hasemani JK, Mahleri J, Aranyi C (1998) Tumor induction in F344/N rats and B6C3F1 mice following inhalation exposure to ethylbenzene. Toxicol Lett 99:23–32

Chapman PM (2000) Whole effluent toxicity testing-usefulness, level of protection, and risk assessment. Environ Toxicol Chem 19:3–13. https://doi.org/10.1002/etc.5620190102

Cheshenko K, Pakdel F, Segner H, Kah O, Eggen RIL (2008) Interference of endocrine disrupting chemicals with aromatase CYP19 expression or activity, and consequences for reproduction of teleost fish. Gen Comp Endocrinol 155:31–62. https://doi.org/10.1016/j.ygcen.2007.03.005

Chiou CT, McGroddy SE, Kile DE (1998) Partition characteristics of polycyclic aromatic hydrocarbons on soils and sediments. Environ Sci Technol 32:264–269. https://doi.org/10.1021/es970614c

Cirelli DP (1978) Patterns of pentachlorophenol usage in the United States of America – an overview. In: Rao KR (ed) Pentachlorophenol. Springer, Boston, pp 13–18. https://doi.org/10.1007/978-1-4615-8948-8_2

Clarkson TW (1993) Molecular and ionic mimicry of toxic metals. Annu Rev Pharmacol Toxicol 33:545–571. https://doi.org/10.1146/annurev.pa.33.040193.002553

Collier TK, Anulacion BF, Arkoosh MR, Dietrich JP, Incardona JP, Johnson LL, Ylitalo GM, Myers MS (2014) Effects on fish of polycyclic aromatic hydrocarbons (PAHs) and naphthenic acid exposures. In: Tierney KB, Farrell AP, Brauner CJ (eds) Organic chemical toxicology of fishes, fish physiology. Academic Press, Cambridge, pp 195–256. https://doi.org/10.1016/b978-0-12-398254-4.00022-4

Connolly CA, Walter LM, Baadsgaard H, Longstaffe FJ (1990) Origin and evolution of formation waters, Alberta Basin, Western Canada sedimentary Basin. I. Chemistry. Appl Geochem 5:375–395. https://doi.org/10.1016/0883-2927(90)90016-x

Cusimano RF, Brakke DF, Chapman GA (1986) Effects of pH on the toxicities of cadmium, copper, and zinc to steelhead trout (Salmo gairdneri). Can J Fish Aquat Sci 43:1497–1503. https://doi.org/10.1139/f86-187

Dalton TP, Puga A, Shertzer HG (2002) Induction of cellular oxidative stress by aryl hydrocarbon receptor activation. Chem Biol Interact 141:77–95

Davies PH, Goettl JP, Sinley JR, Smith NF (1976) Acute and chronic toxicity of lead to rainbow trout Salmo gairdneri, in hard and soft water. Water Res 10:199. https://doi.org/10.1016/0043-1354(76)90128-7

de Maag PG, ten Hulscher DT, van den Heuvel H, Opperhuizen A, Sijm DTHM (1998) Physiochemical properties of polycyclic aromatic hydrocarbons: Aqueous solubilities,

n-octanol/water partition coefficients, and Henry's law constants. Environ Toxicol Chem 17:251–257. https://doi.org/10.1002/etc.5620170216
de Schamphelaere KAC, Heijerick DG, Janssen CR (2002) Refinement and field validation of a biotic ligand model predicting acute copper toxicity to Daphnia magna. Comp Biochem Physiol C Toxicol Pharmacol 133:243–258
DeGraeve GM, Elder RG, Woods DC (1982) Effects of naphthalene and benzene on fathead minnows and rainbow trout. Arch Environ Contam Toxicol 11:487. https://doi.org/10.1007/bf01056076
Department of Environmental Protection (DEP) Pennsylvania (2010) PADEP frac and flowback water analytical data – inorganics. PADEP right-to-know request. Palmerton Group Environmental Consulting Services, New York
Department of Environmental Protection (DEP) Pennsylvania (2016) Technologically enhanced naturally occurring radioactive materials (TENORM) study report. Perma-Fix Environmental Services, Pennsylvania. https://www.dep.pa.gov/Business/Energy/OilandGasPrograms/OilandGasMgmt/Oil-and-Gas-Related-Topics/Pages/Radiation-Protection.aspx. Accessed May 2018
Diamond SA, Oris JT, Guttman SI (1995) Adaptation to fluoranthene exposure in a laboratory population of fathead minnows. Environ Toxicol Chem 14:1393. https://doi.org/10.1002/etc.5620140816
DiGiulio DC, Jackson RB (2016) Impact to underground sources of drinking water and domestic wells from production well stimulation and completion practices in the Pavillion, Wyoming, Field. Environ Sci Technol 50:4524–4536. https://doi.org/10.1021/acs.est.5b04970
Dresel PE, Rose AW (2010) Chemistry and origin of oil and gas well brines in western Pennsylvania: Pennsylvania Geological Survey, 4th ser., Open-File Report OFOG 10–01.0, 48 p. https://www.fwspubs.org/doi/suppl/10.3996/052013-JFWM-033/suppl_file/patnodereference+s3.pdf. Accessed Oct 2017
Dwyer FJ, Mayer FL, Sappington LC, Buckler DR, Bridges CM, Greer IE, Hardesty DK, Henke CE, Ingersoll CG, Kunz JL, Whites DW, Augspurger T, Mount DR, Hattala K, Neuderfer GN (2005) Assessing contaminant sensitivity of endangered and threatened aquatic species: part I. Acute toxicity of five chemicals. Arch Environ Contam Toxicol 48:143–154. https://doi.org/10.1007/s00244-003-3038-1
Eastmond DA, Booth GM, Lee ML (1984) Toxicity, accumulation, and elimination of polycyclic aromatic sulfur heterocycles in Daphnia magna. Arch Environ Contam Toxicol 13:105–111
Edsall CC (1991) Acute toxicities to larval rainbow trout of representative compounds detected in Great Lakes fish. Bull Environ Contam Toxicol 46:173. https://doi.org/10.1007/bf01691933
Entrekin S, Evans-White M, Johnson B, Hagenbuch E (2011) Rapid expansion of natural gas development poses a threat to surface waters. Front Ecol Environ 9:503–511. https://doi.org/10.1890/110053
Eom IC, Rast C, Veber AM, Vasseur P (2007) Ecotoxicity of a polycyclic aromatic hydrocarbon (PAH)-contaminated soil. Ecotoxicol Environ Saf 67:190–205. https://doi.org/10.1016/j.ecoenv.2006.12.020
Erickson RJ, Benoit DA, Mattson VR (1996) The effects of water chemistry on the toxicity of copper to fathead minnows. Environ Toxicol Chem 15:181. https://doi.org/10.1002/etc.5620150217
Evans DH, Piermarini PM, Choe KP (2005) The multifunctional fish gill: dominant site of gas exchange, osmoregulation, acid-base regulation, and excretion of nitrogenous waste. Physiol Rev 85:97–177. https://doi.org/10.1152/physrev.00050.2003
Ewell WS, Gorsuch JW, Kringle RO (1986) Simultaneous evaluation of the acute effects of chemicals on seven aquatic species. Environ Toxicol Chem 5:831. https://doi.org/10.1002/etc.5620050908

Faber A, Annevelink MPJA, Gilissen HK, Schot PP, van Rijswick M, de Voogt P, van Wezel AP (2017) How to adapt chemical risk assessment for unconventional hydrocarbon extraction related to the water system. Rev Environ Contam Toxicol 246:1–32. https://doi.org/10.1007/398_2017_10

Faber A, Annevelink MPJA, Schot PP, Baken KA, Schriks M, Emke E, de Voogt P, van Wezel AP (2019) Chemical and bioassay assessment of waters related to hydraulic fracturing at a tight gas production site. Sci Total Environ 690:636–646. https://doi.org/10.1016/j.scitotenv.2019.06.354

Fang Q, Shi X, Zhang L, Wang Q, Wang X, Guo Y, Zhou B (2015) Effect of titanium dioxide nanoparticles on the bioavailability, metabolism, and toxicity of pentachlorophenol in zebrafish larvae. J Hazard Mater 283:897–904. https://doi.org/10.1016/j.jhazmat.2014.10.039

Feldmannová M, Hilscherová K, Marsálek B, Bláha L (2006) Effects of N-heterocyclic polyaromatic hydrocarbons on survival, reproduction, and biochemical parameters in Daphnia magna. Environ Toxicol 21:425–431. https://doi.org/10.1002/tox.20198

Ferrar KJ, Michanowicz DR, Christen CL, Mulcahy N, Malone SL, Sharma RK (2013) Assessment of effluent contaminants from three facilities discharging Marcellus shale wastewater to surface waters in Pennsylvania. Environ Sci Technol 27:3472–3481. https://doi.org/10.1021/es301411q

Ferrer I, Thurman EM (2015) Analysis of hydraulic fracturing additives by LC/Q-TOF-MS. Anal Bioanal Chem 407:6417–6428. https://doi.org/10.1007/s00216-015-8780-5

Fisher D, Yonkos L, Ziegler G, Friedel E, Burton D (2014) Acute and chronic toxicity of selected disinfection byproducts to Daphnia magna, Cyprinodon variegatus, and Isochrysis galbana. Water Res 55:233–244. https://doi.org/10.1016/j.watres.2014.01.056

Folkerts EJ, Blewett TA, He Y, Goss GG (2017a) Alterations to juvenile zebrafish (Danio rerio) swim performance after acute embryonic exposure to sub-lethal exposures of hydraulic fracturing flowback and produced water. Aquat Toxicol 193:50–59. https://doi.org/10.1016/j.aquatox.2017.10.003

Folkerts EJ, Blewett TA, He Y, Goss GG (2017b) Cardio-respirometry disruption in zebrafish (Danio rerio) embryos exposed to hydraulic fracturing flowback and produced water. Environ Pollut 231:1477–1487. https://doi.org/10.1016/j.envpol.2017.09.011

Folkerts EJ, Blewett TA, Delompré P, Mehler WT, Flynn SL, Sun C, Zhang Y, Martin JW, Alessi DS, Goss GG (2019) Toxicity in aquatic model species exposed to a temporal series of three different flowback and produced water samples collected from a horizontal hydraulically fractured well. Ecotoxicol Environ Saf 180:600–609. https://doi.org/10.1016/j.ecoenv.2019.05.054

Foster WG, McMahon A, Younglai EV, Jarrell JF, Lecavalier P (1995) Alterations in circulating ovarian steroids in hexachlorobenzene-exposed monkeys. Reprod Toxicol 9:541–548

Gagnon GA, Krkosek W, Anderson L, McBean E, Mohseni M, Bazri M, Mauro I (2016) Impacts of hydraulic fracturing on water quality: a review of literature, regulatory frameworks and an analysis of information gaps. Environ Rev 24:122–131. https://doi.org/10.1139/er-2015-0043

García MA, Peña D, Alvarez L, Cocca C, Pontillo C, Bergoc R, de Pisarev DK, Randi A (2010) Hexachlorobenzene induces cell proliferation and IGF-I signaling pathway in an estrogen receptor alpha-dependent manner in MCF-7 breast cancer cell line. Toxicol Lett 192:195–205. https://doi.org/10.1016/j.toxlet.2009.10.026

Gauthier PT, Norwood WP, Prepas EE, Pyle GG (2014) Metal-PAH mixtures in the aquatic environment: a review of co-toxic mechanisms leading to more-than-additive outcomes. Aquat Toxicol 154:253–269. https://doi.org/10.1016/j.aquatox.2014.05.026

Gehman J, Thompson D, Alessi D, Allen D, Goss G (2016) Comparative analysis of hydraulic fracturing wastewater practices in unconventional shale development: newspaper coverage of stakeholder concerns and social license to operate. Sustainability 8:1–23. https://doi.org/10.3390/su8090912

Geiger JG, Buikema AL Jr (1982) Hydrocarbons depress growth and reproduction of Daphnia pulex (Cladocera). Can J Fish Aquat Sci 39:830–836. https://doi.org/10.1139/f82-113

Gesto M, Tintos A, Soengas JL, Míguez JM (2006) Effects of acute and prolonged naphthalene exposure on brain monoaminergic neurotransmitters in rainbow trout (Oncorhynchus mykiss). Comp Biochem Physiol C Toxicol Pharmacol 144:173–183. https://doi.org/10.1016/j.cbpc.2006.08.002

Gesto M, Tintos A, Soengas JL, Míguez JM (2009) beta-Naphthoflavone and benzo(a)pyrene alter dopaminergic, noradrenergic, and serotonergic systems in brain and pituitary of rainbow trout (Oncorhynchus mykiss). Ecotoxicol Environ Saf 72:191–198. https://doi.org/10.1016/j.ecoenv.2008.03.007

Ghazy M, Habashy MM, Kossa FI (2009) Effects of salinity on survival, growth and reproduction of the water flea, Daphnia magna. Nat Sci 7:28–41

Goodnight CJ (1942) Toxicity of sodium pentachlorophenate and pentachlorophenol to fish. Ind Eng Chem 34:868–872. https://doi.org/10.1021/ie50391a022

Goss G, Gilmour K, Hawkings G, Brumbach JH, Huynh M, Galvez F (2011) Mechanism of sodium uptake in PNA negative MR cells from rainbow trout, Oncorhynchus mykiss as revealed by silver and copper inhibition. Comp Biochem Physiol Part A Mol Integr Physiol 159:234–241. https://doi.org/10.1016/j.cbpa.2011.02.025

Goss G, Alessi D, Allen D, Gehman J, Brisbois J, Kletke S, Sharak AZ, Notte C, Thompson DY, Hong K, Prosser C (2015) Unconventional wastewater management: a comparative review and analysis of hydraulic fracturing wastewater management practices across four North American basins. Canadian Water Network, Waterloo. http://cwn-rce.ca/project/unconventional-wastewater-management-a-comparative-review-and-analysis-of-hydraulic-fracturing-wastewater-management-practices-across-four-north-american-basins/

Gregoraszczuk EŁ, Ptak A, Rak-Mardyła A, Falandysz J (2011) Differential accumulation of HCBz and PeCBz in porcine ovarian follicles and their opposing actions on steroid secretion and CYP11, CYP17, 17β-HSD and CYP19 protein expression. A tissue culture approach. Reprod Toxicol 31:494–499. https://doi.org/10.1016/j.reprotox.2011.01.006

Grosell M (2012) Copper, in: Wood, C.M., Farrell, A.P., Brauner, C.J. (Eds.) Homeostasis and toxicology of essential metals. Academic PressCambridge, pp. 54–135. doi: https://doi.org/10.1016/S1546-5098(11)31002-3

Grosell M, Wood CM (2002) Copper uptake across rainbow trout gills: mechanisms of apical entry. J Exp Biol 205:1179–1188

Haluszczak LO, Rose AW, Kump LR (2013) Geochemical evaluation of flowback brine from Marcellus gas wells in Pennsylvania, USA. Appl Geochem 28:55–61. https://doi.org/10.1016/j.apgeochem.2012.10.002

Han J, Won EJ, Hwang DS, Shin KH, Lee YS, Leung KMY, Lee SJ, Lee JS (2014) Crude oil exposure results in oxidative stress-mediated dysfunctional development and reproduction in the copepod Tigriopus japonicus and modulates expression of cytochrome P450 (CYP) genes. Aquat Toxicol 152:308–317. https://doi.org/10.1016/j.aquatox.2014.04.027

Hanson R, Dodoo DK, Essumang DK, Blay J, Yankson K (2007) The effect of some selected pesticides on the growth and reproduction of fresh water Oreochromis niloticus, Chrysichthys nigrodigitatus and Clarias gariepinus. Bull Environ Contam Toxicol 79:544–547. https://doi.org/10.1007/s00128-007-9279-3

Hayes T (2009) Sampling and analysis of water streams associated with the development of Marcellus shale gas. Report Characterization of Marcellus shale flowback water 523 by Gas Technology Institute, Des Plaines, IL, for the Marcellus Shale Coalition

He Y, Flynn SL, Folkerts EJ, Zhang Y, Ruan D, Alessi DS, Martin JW, Goss GG (2017a) Chemical and toxicological characterizations of hydraulic fracturing flowback and produced water. Water Res 114:78–87. https://doi.org/10.1016/j.watres.2017.02.027

He Y, Folkerts EJ, Zhang Y, Martin JW, Alessi DS, Goss GG (2017b) Effects on biotransformation, oxidative stress, and endocrine disruption in rainbow trout (Oncorhynchus mykiss) exposed to hydraulic fracturing flowback and produced water. Environ Sci Technol 51:940–947. https://doi.org/10.1021/acs.est.6b04695

He Y, Sun C, Zhang Y, Folkerts EJ, Martin JW, Goss GG (2018) Developmental toxicity of the organic fraction from hydraulic fracturing flowback and produced waters to early life stages of zebrafish (Danio rerio). Environ Sci Technol 52:3820–3830. https://doi.org/10.1021/acs.est.7b06557

Heintz RA, Short JW, Rice SD (1999) Sensitivity of fish embryos to weathered crude oil: Part II. Increased mortality of pink salmon (Oncorhynchus gorbuscha) embryos incubating downstream from weathered Exxon Valdez crude oil. Environ Toxicol Chem 18:494–503. https://doi.org/10.1002/etc.5620180318

Hermens J, Canton H, Steyger N, Wegman R (1984) Joint effects of a mixture of 14 chemicals on mortality and inhibition of reproduction of Daphnia magna. Aquat Toxicol 5:315. https://doi.org/10.1016/0166-445x(84)90012-2

Ho TY, Witz G (1997) Increased gene expression in human promyeloid leukemia cells exposed to trans,trans-muconaldehyde, a hematotoxic benzene metabolite. Carcinogenesis 18:739–744

Hogstrand C (2012) Zinc. In: Wood CM, Farrell AP, Brauner CJ (eds) Homeostasis and toxicology of essential metals. Academic Press, Cambridge, pp 136–231. https://doi.org/10.1016/s1546-5098(11)31032-1

Hogstrand C, Wilson RW, Polgar D, Wood CM (1994) Effects of zinc on the kinetics of branchial calcium uptake in freshwater rainbow trout during adaptation to waterborne zinc. J Exp Biol 186:55–73

Holcombe GW, Phipps GL, Fiandt JT (1983) Toxicity of selected priority pollutants to various aquatic organisms. Ecotoxicol Environ Saf 7:400. https://doi.org/10.1016/0147-6513(83)90005-2

Hollis L, McGeer JC, McDonald DG (2000) Protective effects of calcium against chronic waterborne cadmium exposure to juvenile rainbow trout. Environ Toxicol Chem 19:2725. https://doi.org/10.1002/etc.5620191117

Howarth RS, Sprague JB (1978) Copper lethality to rainbow trout in waters of various hardness and pH. Water Res 12:455. https://doi.org/10.1016/0043-1354(78)90152-5

Frac Focus Chemical Disclosure Registry (n.d.). http://fracfocus.org/chemical-use/what-chemicals-are-used. Accessed Sept 2018

Hughes GM (1972) Morphometrics of fish gills. Respir Physiol 14:1–25. https://doi.org/10.1016/0034-5687(72)90014-x

Incardona JP, Scholz NL (2016) The influence of heart developmental anatomy on cardiotoxicity-based adverse outcome pathways in fish. Aquat Toxicol 177:515–525

Incardona JP, Collier TK, Scholz NL (2004) Defects in cardiac function precede morphological abnormalities in fish embryos exposed to polycyclic aromatic hydrocarbons. Toxicol Appl Pharmacol 196:191–205

Incardona JP, Carls MG, Day HL, Sloan CA, Bolton JL, Collier TK, Scholz NL (2009) Cardiac arrhythmia is the primary response of embryonic Pacific herring (Clupea pallasi) exposed to crude oil during weathering. Environ Sci Technol 43:201–207

Incardona JP, Linbo TL, Scholz NL (2011) Cardiac toxicity of 5-ring polycyclic aromatic hydrocarbons is differentially dependent on the aryl hydrocarbon receptor 2 isoform during zebrafish development. Toxicol Appl Pharmacol 257:242–249. https://doi.org/10.1016/j.taap.2011.09.010

Incardona JP, Swarts TL, Edmunds RC, Linbo TL, Aquilina-Beck A, Sloan CA, Gardner LD, Block BA, Scholz NL (2013) Exxon valdez to deepwater horizon: comparable toxicity of both crude oils to fish early life stages. Aquat Toxicol 142–143:303–316. https://doi.org/10.1016/j.aquatox.2013.08.011

Incardona JP, Gardner LD, Linbo TL (2014) Deepwater Horizon crude oil impacts the developing hearts of large predatory pelagic fish. Proc Natl Acad Sci 111:E1510. https://doi.org/10.1073/pnas.1320950111

Itoh N, Shinya M (1994) Seasonal evolution of bromomethanes from coralline algae (Corallinaceae) and its effect on atmospheric ozone. Mar Chem 45:95–103. https://doi.org/10.1016/0304-4203(94)90094-9

James MO (1989) Biotransformation and disposition of PAH in aquatic invertebrates. In: Varanasi U (ed) Metabolism of polycyclic aromatic hydrocarbons in the aquatic environment. CRC Press, Boca Raton, pp 66–91

Johnson WW, Finley MT (1980) Handbook of acute toxicity of chemicals to fish and aquatic invertebrates: Summaries of toxicity tests conducted at Columbia National Fisheries Research Laboratory. United States Fish and Wildlife Service. https://pubs.er.usgs.gov/publication/rp137

Johnson JL, Stalling DL, Hogan JW (1974) Hexachlorobenzene (HCB) residues in fish. Bull Environ Contam Toxicol 11:393–398. https://doi.org/10.1007/bf01685293

Johnson LL, Stehr CM, Olson OP, Myers MS, Pierce SM, Wigren CA, McCain BB, Varanasi U (1993) Chemical contaminants and hepatic lesions in winter flounder (Pleuronectes americanus) from the northeast coast of the United States. Environ Sci Technol 27:2759–2771. https://doi.org/10.1021/es00049a015

Jones NF, Pejchar L, Kiesecker JM (2015) The energy footprint: how oil, natural gas, and wind energy affect land for biodiversity and the flow of ecosystem services. Bioscience 65:290–301. https://doi.org/10.1093/biosci/biu224

Juhasz AL, Naidu R (2000) Bioremediation of high molecular weight polycyclic aromatic hydrocarbons: a review of the microbial degradation of benzo[a]pyrene. Int Biodeterior Biodegradation 45:57–88. https://doi.org/10.1016/s0964-8305(00)00052-4

Jung J, Ishida K, Nishihara T (2004) Anti-estrogenic activity of fifty chemicals evaluated by in vitro assays. Life Sci 74:3065–3074. https://doi.org/10.1016/j.lfs.2003.10.030

Jung JH, Hicken CE, Boyd D, Anulacion BF, Carls MG, Shim WJ, Incardona JP (2013) Geologically distinct crude oils cause a common cardiotoxicity syndrome in developing zebrafish. Chemosphere 91:1146–1155. https://doi.org/10.1016/j.chemosphere.2013.01.019

Kahrilas GA, Blotevogel J, Corrin ER, Borch T (2016) Downhole transformation of the hydraulic fracturing fluid biocide glutaraldehyde: implications for flowback and produced water quality. Environ Sci Technol 50:11414–11423. https://doi.org/10.1021/acs.est.6b02881

Kappus H (1987) Oxidative stress in chemical toxicity. Arch Toxicol 60:144–149

Kassotis CD, Nagel SC, Stapleton HM (2018a) Unconventional oil and gas chemicals and wastewater-impacted water samples promote adipogenesis via PPARγ-dependent and independent mechanisms in 3T3-L1 cells. Sci Total Environ 640–641:1601–1610. https://doi.org/10.1016/j.scitotenv.2018.05.030

Kassotis CD, Vu DC, Vo PH, Lin CH, Cornelius-Green JN, Patton S, Nagel SC (2018b) Endocrine-disrupting activities and organic contaminants associated with oil and gas operations in Wyoming groundwater. Arch Environ Contam Toxicol 75:247–258. https://doi.org/10.1007/s00244-018-0521-2

Katukam V, Kulakarni M, Syed R, Alharbi K, Naik J (2012) Effect of benzene exposure on fertility of male workers employed in bulk drug industries. Genet Test Mol Biomarkers 16:592–597. https://doi.org/10.1089/gtmb.2011.0241

Kelly JD, Dutchuk M, Hendricks JD, Williams DE (1993) Hepatocarcinogenic potency of mixed and pure enantiomers of trans-7,8-dihydrobenzo[a]pyrene-7,8-diol in trout. Cancer Lett 68:225–229

Khursigara AJ, Perrichon P, Bautista NM, Burggren WW, Esbaugh AJ (2017) Cardiac function and survival are affected by crude oil in larval red drum, Sciaenops ocellatus. Sci Total Environ 579:797–804. https://doi.org/10.1016/j.scitotenv.2016.11.026

Köhler A, Wahl E, Söffker K (2002) Functional and morphological changes of lysosomes as prognostic biomarkers of toxic liver injury in a marine flatfish (Platichthys flesus (L.)). Environ Toxicol Chem 21:2434–2444

Kolachana P, Subrahmanyam VV, Meyer KB, Zhang L (1993) Benzene and its phenolic metabolites produce oxidative DNA damage in HL60 cells in vitro and in the bone marrow in vivo. Cancer Res 53:1023–1026

Kondash A, Vengosh A (2015) Water footprint of hydraulic fracturing. Environ Sci Technol Lett 2:276–280. https://doi.org/10.1021/acs.estlett.5b00211

Kulander CS (2013) Shale oil and gas state regulatory issues and trends. Case West Reserv Law Rev 63:1101–1141

Lampi MA, Gurska J, McDonald K (2006) Photoinduced toxicity of polycyclic aromatic hydrocarbons to Daphnia magna: ultraviolet-mediated effects and the toxicity of polycyclic aromatic hydrocarbon photoproducts. Environ Toxicol Chem 25:1079. https://doi.org/10.1897/05-276r.1

Lauer NE, Harkness JS, Vengosh A (2016) Brine spills associated with unconventional oil development in North Dakota. Environ Sci Technol 50:5389–5397. https://doi.org/10.1021/acs.est.5b06349

Le Page Y, Vosges M, Servili A, Brion F, Kah O (2011) Neuroendocrine effects of endocrine disruptors in teleost fish. J Toxicol Environ Health B Crit Rev 14:370–386. https://doi.org/10.1080/10937404.2011.578558

LeBlanc GA (1980) Acute toxicity of priority pollutants to water flea (Daphnia magna). Bull Environ Contam Toxicol 24:684. https://doi.org/10.1007/bf01608174

Lester Y, Ferrer I, Thurman EM, Sitterley KA, Korak JA, Aiken G, Linden KG (2015) Characterization of hydraulic fracturing flowback water in Colorado: implications for water treatment. Sci Total Environ 512–513:637–644. https://doi.org/10.1016/j.scitotenv.2015.01.043

Li M, Yao H, Stasiuk LD, Fowler MG, Larter SR (1997) Effect of maturity and petroleum expulsion on pyrrolic nitrogen compound yields and distributions in Duvernay Formation petroleum source rocks in central Alberta, Canada. Org Geochem 26:731–744. https://doi.org/10.1016/s0146-6380(97)00053-3

Li J, Ma M, Wang Z (2008) A two-hybrid yeast assay to quantify the effects of xenobiotics on retinoid X receptor-mediated gene expression. Toxicol Lett 176:198–206. https://doi.org/10.1016/j.toxlet.2007.11.006

Lima ALC, Farrington JW, Reddy CM (2005) Combustion-derived polycyclic aromatic hydrocarbons in the environment – a review. Environ Forensic 6:109–131. https://doi.org/10.1080/15275920590952739

Liu J, Goyer RA, Waalkes MP (2008) Toxic effects of metals. In: Klaassen CD (ed) Casarett and Doull's toxicology: the basic science of poisons. McGraw-Hill, New York, pp 931–980

Livingstone DR (1998) The fate of organic xenobiotics in aquatic ecosystems: quantitative and qualitative differences in biotransformation by invertebrates and fish. Comp Biochem Physiol Part A Mol Integr Physiol 120:43–49

Lores EM, Pennock JR (1998) The effect of salinity on binding of Cd, Cr, Cu and Zn to dissolved organic matter. Chemosphere 37:861–874. https://doi.org/10.1016/s0045-6535(98)00090-3

Lushchak VI (2011) Environmentally induced oxidative stress in aquatic animals. Aquat Toxicol 101:13–30. https://doi.org/10.1016/j.aquatox.2010.10.006

Ma Y, Lei YD, Xiao H, Wania F, Wang W (2010) Critical review and recommended values for the physio-chemical property data of 15 polycyclic aromatic hydrocarbons at 25°C. J Chem Eng Data 55:819–825. https://doi.org/10.1021/je900477x

Mackay D, McAuliffe CD (1988) Fate of hydrocarbons discharged at sea. Oil Chem Pollut 5:1–20. https://doi.org/10.1016/s0269-8579(89)80002-4

Mackay D, Shiu WY (1981) A critical review of Henry's Law constants for chemicals of environmental interest. J Phys Chem Ref Data 10:1175–1199

Mager EM (2012) Lead. In: Wood CM, Farrell AP, Brauner CJ (eds) Homeostasis and toxicology of non-essential metals. Academic Press, Cambridge, pp 186–236

Maguire-Boyle SJ, Barron AR (2014) Organic compounds in produced waters from shale gas wells. Environ Sci Process Impacts 16:2237–2248. https://doi.org/10.1039/c4em00376d

Maronpot RR (1987) Ovarian toxicity and carcinogenicity in eight recent National Toxicology Program studies. Environ Health Perspect 73:125–130

Martínez-Jerónimo F, Martínez-Jerónimo L (2007) Chronic effect of NaCl salinity on a freshwater strain of Daphnia magna Straus (Crustacea: Cladocera): a demographic study. Ecotoxicol Environ Saf 67:411–416. https://doi.org/10.1016/j.ecoenv.2006.08.009

Marty GD, Hinton DE, Short JW, Heintz RA, Rice SD, Dambach DM, Willits NH, Stegeman JJ (1997) Ascites, premature emergence, increased gonadal cell apoptosis, and cytochrome P4501A induction in pink salmon larvae continuously exposed to oil-contaminated gravel during development. Can J Zool 75:989–1007. https://doi.org/10.1139/z97-120

Mayer P, Reichenberg F (2006) Can highly hydrophobic organic substances cause aquatic baseline toxicity and can they contribute to mixture toxicity? Environ Toxicol Chem 25:2639–2644

McGeer JC, Niyogi S, Smith DS (2012) Cadmium, in: Wood CM, Farrell AP, Brauner CJ (eds) Homeostasis and toxicology of non-essential metals. Academic Press, Cambridge, pp 126–185.

McKim JM, Schmieder PK, Carlson RW, Hunt EP, Niemi GJ (1987) Use of respiratory-cardiovascular responses of rainbow trout (Salmo gairdneri) in identifying acute toxicity syndromes in fish: Part 1. pentachlorophenol, 2,4-dinitrophenol, tricaine methane sulfonate and 1-octanol. Environ Toxicol Chem 6:295–312. https://doi.org/10.1002/etc.5620060407

Meador JP, Stein JE, Reichert WL, Varanasi U (1995) Bioaccumulation of polycyclic aromatic hydrocarbons by marine organisms. In: Ware GW (ed) Reviews of environmental contamination and toxicology. Springer, New York, pp 79–165. https://doi.org/10.1007/978-1-4612-2542-3_4

Millemann RE, Birge WJ, Black JA (1984) Comparative acute toxicity to aquatic organisms of components of coal-derived synthetic fuels. Trans Am Fish Soc 113:74–85. https://doi.org/10.1577/1548-8659(1984)113<74:CATTAO>2.0.CO;2

Mount DI (1968) Chronic toxicity of copper to fathead minnows (Pimephales promelas, Rafinesque). Water Res 2:215. https://doi.org/10.1016/0043-1354(68)90027-4

Mount DI, Stephan CE (1969) Chronic toxicity of copper to the fathead minnow (Pimephales promelas) in soft water. J Fish Res Board Can 26:2449. https://doi.org/10.1139/f69-236

Mount DR, Gulley DD, Hockett JR, Garrison TD, Evans JM (1997) Statistical models to predict the toxicity of major ions to *Ceriodaphnia dubia*, *Daphnia magna*, and *Pimephales promelas* (fathead minnows). Environ Toxicol Chem 16:2009–2019. https://doi.org/10.1002/etc.5620161005

Mundy LJ, Jones SP, Crump D, Hervé JC, Konstantinov A, Utley F, Potter D, Kennedy SW (2010) Highly purified hexachlorobenzene induces cytochrome P4501A in primary cultures of chicken embryo hepatocytes. Toxicol Appl Pharmacol 248:185–193. https://doi.org/10.1016/j.taap.2010.07.022

Mundy LJ, Crump D, Jones SP, Konstantinov A, Utley F, Potter D, Kennedy SW (2012) Induction of cytochrome P4501A by highly purified hexachlorobenzene in primary cultures of ring-necked pheasant and Japanese quail embryo hepatocytes. Comp Biochem Physiol C Toxicol Pharmacol 155:498–505. https://doi.org/10.1016/j.cbpc.2011.12.006

Muñoz MJ, Tarazona JV (1993) Synergistic effect of two- and four-component combinations of the polycyclic aromatic hydrocarbons: phenanthrene, anthracene, naphthalene and acenaphthene on Daphnia magna. Bull Environ Contam Toxicol 50:363–368

Munson AE, Sain LE, Sanders VM, Kauffmann BM, White KL, Page DG, Barnes DW, Borzelleca JF (1982) Toxicology of organic drinking water contaminants: trichloromethane, bromodichloromethane, dibromochloromethane and tribromomethane. Environ Health Perspect 46:117–126. https://doi.org/10.1289/ehp.8246117

Nakai N, Murata M, Nagahama M, Hirase T, Tanaka M, Fujikawa T, Nakao N, Nakashima K, Kawanishi S (2003) Oxidative DNA damage induced by toluene is involved in its male reproductive toxicity. Free Radic Res 37:69–76

Nakajima T (1997) Cytochrome P450 isoforms and the metabolism of volatile hydrocarbons of low relative molecular mass. J Occup Health 39:83–91. https://doi.org/10.1539/joh.39.83

Navas JM, Segner H (2001) Estrogen-mediated suppression of cytochrome P4501A (CYP1A) expression in rainbow trout hepatocytes: role of estrogen receptor. Chem Biol Interact 138:285–298

Neff JM, Ostazeski S, Gardiner W, Stejskal I (2000) Effects of weathering on the toxicity of three offshore australian crude oils and a diesel fuel to marine animals. Environ Toxicol Chem 19:1809–1821. https://doi.org/10.1002/etc.5620190715

Neff J, Lee K, DeBlois EM (2011) Produced water: overview of composition, fates, and effects. In: Lee K, Neff J (eds) Produced water: environmental risks and advances in mitigation technologies. Springer, New York, pp 3–54

Nelson PH, Santus SL (2011) Gas, oil, and water production from Wattenberg field in the Denver basin, Colorado. U.S. Geological Survey Open-File Report. https://pubs.er.usgs.gov/publication/ofr20111175

Nelson PO, Ratcliff AK, Schuytema GS (1984) Toxicity of cadmium in water and sediment slurries to Daphnia magna. Environ Toxicol Chem 3:293. https://doi.org/10.1002/etc.5620030212

Nelson D, Heuer RM, Cox GK, Stieglitz JD, Hoenig R, Mager EM, Benetti DD, Grosell M, Crossley DA (2016) Effects of crude oil on in situ cardiac function in young adult mahi-mahi (Coryphaena hippurus). Aquat Toxicol 180:274–281. https://doi.org/10.1016/j.aquatox.2016.10.012

New York State Department of Environmental Conservation (NYSDEC) (2015) Final supplemental generic environmental impact statement on the oil, gas and solution mining regulatory program. Regulatory program for horizontal drilling and high-volume hydraulic fracturing to develop the marcellus shale and other low-permeability gas reservoirs. Volume 1: final supplemental generic environmental impact statement. Volume 2: response to comments. NYSDEC, NY. http://www.dec.ny.gov/docs/materials_minerals_pdf/fsgeis2015.pdf. Accessed Oct 2017

Nicholls EL, Russell AP (1990) Paleobiogeography of the cretaceous Western Interior Seaway of North America: the vertebrate evidence. Palaeogeogr Palaeoclimatol Palaeoecol 79:149–169

Niyogi S, Wood CM (2004) Biotic ligand model, a flexible tool for developing site-specific water quality guidelines for metals. Environ Sci Technol 38:6177–6192

Orem W, Tatu C, Varonka M, Lerch H, Bates A (2014) Organic substances in produced and formation water from unconventional natural gas extraction in coal and shale. Int J Coal Geol 126:20–31

Orton F, Lutz I, Kloas W, Routledge EJ (2009) Endocrine disrupting effects of herbicides and pentachlorophenol: in vitro and in vivo evidence. Environ Sci Technol 43:2144–2150

Park EJ, Jo HJ, Jung J (2009) Combined effects of pH, hardness and dissolved organic carbon on acute metal toxicity to Daphnia magna. J Ind Eng Chem 15:82. https://doi.org/10.1016/j.jiec.2008.08.007

Parkhurst BR, Bradshaw AS, Forte JL (1981) The chronic toxicity to Daphnia magna of acridine, a representative azaarene present in synthetic fossil fuel products and wastewaters. Environ Pollut Series Ecol Biol 24:21–30. https://doi.org/10.1016/0143-1471(81)90119-7

Parthasarathy N, Pelletier M, Buffle J (2008) Transport of lipophilic ligands through permeation liquid membrane in relation to natural water analysis. J Membr Sci 309:182–188. https://doi.org/10.1016/j.memsci.2007.10.015

Parthasarathy N, Pelletier M, Buffle J (2010) Transport of lipophilic metal complexes through permeation liquid membrane, in relation to natural water analysis: Cu(II)–8-hydroxyquinoline complex as a model compound. J Membr Sci 355:78–84. https://doi.org/10.1016/j.memsci.2010.03.008

Patterson LA, Konschnik KE, Wiseman H, Fargione J, Maloney KO, Kiesecker J, Nicot JP, Baruch-Mordo S, Entrekin S, Trainor A, Saiers JE (2017) Unconventional oil and gas spills: risks, mitigation priorities, and state reporting requirements. Environ Sci Technol 51:2563–2573. https://doi.org/10.1021/acs.est.6b05749

Pearson JG, Glennon JP, Barkley JJ (1979) An approach to the toxicological evaluation of a complex industrial wastewater. In: Marking LL, Kimerle RA (eds) Aquatic toxicology. American Society for Testing and Materials, Pennsylvania, pp 284–301. https://doi.org/10.1520/stp34894s

Peña D, Pontillo C, García MA, Cocca C, Alvarez L, Chiappini F, Bourguignon N, Frahm I, Bergoc R, de Pisarev DK, Randi A (2012) Alterations in c-Src/HER1 and estrogen receptor α signaling pathways in mammary gland and tumors of hexachlorobenzene-treated rats. Toxicology 293:68–77. https://doi.org/10.1016/j.tox.2011.12.012

Pennsylvania Department of Environmental Protection (2016) PA DEP oil & gas reporting website: waste reports; https://www.paoilandgasreporting.state.pa.us/publicreports/Modules/Welcome/ProdWasteReports.aspx. Accessed Oct 2018

Peterson DR (1994) Calculating the aquatic toxicity of hydrocarbon mixtures. Chemosphere 29:2493. https://doi.org/10.1016/0045-6535(94)90052-3

Phipps GL, Holcombe GW, Fiandt JT (1981) Acute toxicity of phenol and substituted phenols to the fathead minnow. Bull Environ Contam Toxicol 26:585–593

Pickering QH, Henderson C (1966) Acute toxicity of some important petrochemicals to fish. Water Res 13:623. https://doi.org/10.1016/0043-1354(79)90010-1

Pietsch C, Hollender J, Dorusch F, Burkhardt-Holm P (2014) Cytotoxic effects of pentachlorophenol (PCP) and its metabolite tetrachlorohydroquinone (TCHQ) on liver cells are modulated by antioxidants. Cell Biol Toxicol 30:233–252. https://doi.org/10.1007/s10565-014-9283-4

Pinkney AE, Harshbarger JC (2006) Tumor prevalence in Mummichogs from the Delaware estuary watershed. J Aquat Anim Health 18:244–251. https://doi.org/10.1577/H05-053.1

Playle RC, Dixon DG, Burnison K (1993) Copper and cadmium binding to fish gills: modification by dissolved organic carbon and synthetic ligands. Can J Fish Aquat Sci 50:2667–2677. https://doi.org/10.1139/f93-290

Rahman S, Khan IA, Thomas P (2011) Tryptophan hydroxylase: a target for neuroendocrine disruption. J Toxicol Environ Health B Crit Rev 14:473–494. https://doi.org/10.1080/10937404.2011.578563

Ralston JJ, Kalmbach JA (2018) Regulating under conditions of uncertainty and risk: Lessons learned from state regulation of hydraulic fracturing. Environ Pract 20:1–12. https://doi.org/10.1080/14660466.2018.1444875

Ren S (2002) Predicting three narcosis mechanisms of aquatic toxicity. Toxicol Lett 133:127–139

Reutman SR, LeMasters GK, Knecht EA, Shukla R, Lockey JE, Burroughs GE, Kesner JS (2002) Evidence of reproductive endocrine effects in women with occupational fuel and solvent exposures. Environ Health Perspect 110:805–811

Rice CA (2003) Production waters associated with the Ferron coalbed methane fields, central Utah: chemical and isotopic composition and volumes. Int J Coal Geol 56:141–169. https://doi.org/10.1016/s0166-5162(03)00086-7

Richardson N, Gottlieb M, Krupnick A, Wiseman H (2013) The state of state shale gas regulation. Resources for the future executive summary, Washington. https://www.rff.org/publications/reports/the-state-of-state-shale-gas-regulation/. Accessed Aug 2018

Rivard C, Lavoie D, Lefebvre R, Séjourné S, Lamontagne C, Duchesne M (2014) An overview of Canadian shale gas production and environmental concerns. Int J Coal Geol 126:64–76. https://doi.org/10.1016/j.coal.2013.12.004

Robert J, McGuire CC, Kim F, Nagel SC, Price SJ, Lawrence BP, De Jesús Andino F (2018) Water contaminants associated with unconventional oil and gas extraction cause immunotoxicity to amphibian tadpoles. Toxicol Sci 166(1):39–50. https://doi.org/10.1093/toxsci/kfy179

Rogers JT, Richards JG, Wood CM (2003) Ion regulatory disruption as the acute toxic mechanism for lead in the rainbow trout (Oncorhynchus mykiss). Aquat Toxicol 64:215. https://doi.org/10.1016/s0166-445x(03)00053-5

Roques DE, Overton EB, Henry CB (1994) Using gas chromatography/mass spectroscopy fingerprint analyses to document process and progress of oil degradation. J Environ Qual 23:851–855. https://doi.org/10.2134/jeq1994.00472425002300040034x

Rosenblum J, Thurman EM, Ferrer I, Aiken G, Linden KG (2017a) Organic chemical characterization and mass balance of a hydraulically fractured well: from fracturing fluid to produced water over 405 days. Environ Sci Technol 51:14006–14015. https://doi.org/10.1021/acs.est.7b03362

Rosenblum J, Nelson AW, Ruyle B, Schultz MK, Ryan JN, Linden KG (2017b) Temporal characterization of flowback and produced water quality from a hydraulically fractured oil and gas well. Sci Total Environ 596–597:369–377. https://doi.org/10.1016/j.scitotenv.2017.03.294

Sabljić A, Güsten H, Verhaar H, Hermens J (1995) QSAR modelling of soil sorption. Improvements and systematics of log KOC vs. log KOW correlations. Chemosphere 31:4489–4514. https://doi.org/10.1016/0045-6535(95)00327-5

Schein A, Scott JA, Mos L, Hodson PV (2008) Oil dispersion increases the apparent bioavailability and toxicity of diesel to rainbow trout (Oncorhynchus mykiss). Environ Toxicol Chem 28:595–602. https://doi.org/10.1897/08-315.1

Shaffer DL, Chavez LHA, Ben-Sasson M, Castrillón SRV, Yip NY, Elimelech M (2013) Desalination and reuse of high-salinity shale gas produced water: drivers, technologies, and future directions. Environ Sci Technol 47:9569–9583. https://doi.org/10.1021/es401966e

Short J (2003) Long-term effects of crude oil on developing fish: lessons from the Exxon Valdez oil spill. Energy Sources 25:509–517. https://doi.org/10.1080/00908310390195589

Simonato JD, Fernandes MN, Martinez CBR (2011) Gasoline effects on biotransformation and antioxidant defenses of the freshwater fish Prochilodus lineatus. Ecotoxicology 20:1400–1410. https://doi.org/10.1007/s10646-011-0697-y

Slooff W, Canton JH, Hermens J (1983) Comparison of the susceptibility of 22 freshwater species to 15 chemical compounds. I.(Sub) acute toxicity tests. Aquat Toxicol 4:113. https://doi.org/10.1016/0166-445x(83)90049-8

Smith SB, Savino JF, Blouin MA (1988) Acute toxicity to Daphnia pulex of six classes of chemical compounds potentially hazardous to Great Lakes aquatic biota. J Great Lakes Res 14:394. https://doi.org/10.1016/s0380-1330(88)71572-5

Snarski VM, Olson GF (1982) Chronic toxicity and bioaccumulation of mercuric chloride in the fathead minnow (Pimephales promelas). Aquat Toxicol 2:143. https://doi.org/10.1016/0166-445x(82)90012-1

Sollid J, Nilsson GE (2006) Plasticity of respiratory structures--adaptive remodeling of fish gills induced by ambient oxygen and temperature. Respir Physiol Neurobiol 154:241–251. https://doi.org/10.1016/j.resp.2006.02.006

Song SB, Xu Y, Zhou BS (2006) Effects of hexachlorobenzene on antioxidant status of liver and brain of common carp (Cyprinus carpio). Chemosphere 65:699–706. https://doi.org/10.1016/j.chemosphere.2006.01.033

Soucek DJ, Kennedy AJ (2005) Effects of hardness, chloride, and acclimation on the acute toxicity of sulfate to freshwater invertebrates. Environ Toxicol Chem 24:1204–1210. https://doi.org/10.1897/04-142.1

Spehar RL, Poucher S, Brooke LT, Hansen DJ, Champlin D, Cox DA (1999) Comparative toxicity of fluoranthene to freshwater and saltwater species under fluorescent and ultraviolet light. Arch Environ Contam Toxicol 37:496–502

Stagg RM, Rusin J, McPhail ME, McIntosh AD, Moffat CF, Craft JA (2000) Effects of polycyclic aromatic hydrocarbons on expression of cyp1a in salmon (Salmo salar) following experimental exposure and after the Braer oil spill. Environ Toxicol Chem 19:2797–2805. https://doi.org/10.1002/etc.5620191126

Starek-Świechowicz B, Budziszewska B, Starek A (2017) Hexachlorobenzene as a persistent organic pollutant: toxicity and molecular mechanism of action. Pharmacol Rep 69:1232–1239. https://doi.org/10.1016/j.pharep.2017.06.013

Stringfellow WT, Domen JK, Camarillo MK, Sandelin WL, Borglin S (2014) Physical, chemical, and biological characteristics of compounds used in hydraulic fracturing. J Hazard Mater 275:37–54. https://doi.org/10.1016/j.jhazmat.2014.04.040

Stringfellow WT, Camarillo MK, Domen JK (2017) Identifying chemicals of concern in hydraulic fracturing fluids used for oil production. Environ Pollut 220:413–420. https://doi.org/10.1016/j.envpol.2016.09.082

Suedel BC, Rodgers JH Jr (1996) Toxicity of fluoranthene to Daphnia magna, Hyalella azteca, Chironomus tentans, and Stylaria lacustris in water-only and whole sediment exposures. Bull Environ Contam Toxicol 57:132. https://doi.org/10.1007/s001289900166

Sun Y, Yu H, Zhang J, Yin Y, Shi H, Wang X (2006) Bioaccumulation, depuration and oxidative stress in fish *Carassius auratus* under phenanthrene exposure. Chemosphere 63:1319–1327. https://doi.org/10.1016/j.chemosphere.2005.09.032

Suzuki T, Ide K, Ishida M (2001) Response of MCF-7 human breast cancer cells to some binary mixtures of oestrogenic compounds in-vitro. J Pharm Pharmacol 53:1549–1554

Szebedinszky C, McGeer JC (2001) Effects of chronic Cd exposure via the diet or water on internal organ-specific distribution and subsequent gill Cd uptake kinetics in juvenile rainbow trout. Environ Toxicol Chem 20:597. https://doi.org/10.1002/etc.5620200320

Takayanagi K (2001) Acute toxicity of waterborne Se (IV), Se (VI), Sb (III), and Sb (V) on red seabream (Pagrus major). Bull Environ Contam Toxicol 66:808. https://doi.org/10.1007/s001280080

Tasker TL, Burgos WD, Piotrowski P, Castillo-Meza L, Blewett TA, Ganow KB, Stallworth A, Delompré PLM, Goss GG, Fowler LB, Heuvel JPV, Dorman F, Warner NR (2018) Environmental and human health impacts of spreading oil and gas wastewater on roads. Environ Sci Technol 52:7081–7091. https://doi.org/10.1021/acs.est.8b00716

Taylor LN, McGeer JC, Wood CM (2000) Physiological effects of chronic copper exposure to rainbow trout (Oncorhynchus mykiss) in hard and soft water: evaluation of chronic indicators. Environ Toxicol Chem 19:2298. https://doi.org/10.1002/etc.5620190920

Theodori GL, Luloff AE, Willits FK, Burnett DB (2014) Hydraulic fracturing and the management, disposal, and reuse of frac flowback waters: views from the public in the Marcellus Shale. Energy Res Soc Sci 2:66–74. https://doi.org/10.1016/j.erss.2014.04.007

Thurman EM (1985) Organic geochemistry of natural waters, 1st edn. Martinus Nijhof/ Dr. W. Junk Publishers, Boston

Tišler T, Zagorc J (2002) Acute and chronic toxicity of arsenic to some aquatic organisms. Bull Environ Contam Toxicol 69:421. https://doi.org/10.1007/s00128-002-0079-5

Trucco RG, Engelhardt FR, Stacey B (1983) Toxicity, accumulation and clearance of aromatic hydrocarbons in Daphnia pulex. Environ Pollut Series Ecol Biol 31:191. https://doi.org/10.1016/0143-1471(83)90076-4

U.S. EPA (1978) In-depth studies on health and environmental impacts of selected water pollutants monthly report. Task II: Static acute bioassays and physical chemistry measurements. Contract number 68-01-4646. https://semspub.epa.gov/work/10/100004094.pdf. Accessed Oct 2018

U.S. EPA (1984) Aquatic toxicity tests to characterize the hazard of volatile organic chemicals in water: a toxicity data summary. Parts I and II. Report number 600/S3-84-009. https://nepis.epa.gov/Exe/ZyPDF.cgi/2000H8SM.PDF?Dockey=2000H8SM.PDF. Accessed Nov 2018

U.S. EPA (2015) Atlas resources sample report. U.S. Environmental Protection Agency, Washington, DC. https://www.epa.gov/sites/production/files/2015-05/documents/atlas_response_211-419.pdf. Accessed Mar 2018

U.S. EPA (2016) Hydraulic fracturing for oil and gas: impacts from the hydraulic fracturing water cycle on drinking water resources in the United States (Final Report). U.S. Environmental Protection Agency, Washington, EPA/600/R-16/236F. https://www.epa.gov/sites/production/files/2015-05/documents/hf_spills_report_final_5-12-15_508_km_sb.pdf. Accessed Nov 2018

Valavanidis A, Vlahogianni T, Dassenakis M, Scoullos M (2006) Molecular biomarkers of oxidative stress in aquatic organisms in relation to toxic environmental pollutants. Ecotoxicol Environ Saf 64:178–189. https://doi.org/10.1016/j.ecoenv.2005.03.013

Varanasi U, Stein JE, Nishimoto M (1989) Biotransformation and disposition of polycyclic aromatic hydrocarbons (PAH) in fish. In: Varanasi U (ed) Metabolism of polycyclic aromatic hydrocarbons in the aquatic environment. CRC Press, Boca Raton, pp 93–149

Veith GD, Call DJ, Brooke LT (1983) Estimating the acute toxicity of narcotic industrial chemicals to fathead minnows. In: Bishop WE, Cardwell RD, Heidolph BB (eds) Aquatic toxicology and hazard assessment. American Society for Testing and Materials, Philadelphia, pp 90–97. https://doi.org/10.1520/stp33499s

Vengosh A, Jackson RB, Warner N, Darrah TH, Kondash A (2014) A critical review of the risks to water resources from unconventional shale gas development and hydraulic fracturing in the United States. Environ Sci Technol 48:8334–8348. https://doi.org/10.1021/es405118y

Verma Y, Rana SVS (2009) Endocrinal toxicity of industrial solvents--a mini review. Indian J Exp Biol 47:537–549

Villeneuve DL, Khim JS, Kannan K, Giesy JP (2002) Relative potencies of individual polycyclic aromatic hydrocarbons to induce dioxinlike and estrogenic responses in three cell lines. Environ Toxicol 17:128–137. https://doi.org/10.1002/tox.10041

Wang XC, Zhang YX, Chen RF (2001) Distribution and partitioning of polycyclic aromatic hydrocarbons (PAHs) in different size fractions in sediments from Boston Harbor, United States. Mar Pollut Bull 42:1139–1149

Ward CO, Kuna RA, Snyder NK, Alsaker RD, Coate WB, Craig PH (1985) Subchronic inhalation toxicity of benzene in rats and mice. Am J Ind Med 7:457–473

Webb E, Bushkin-Bedient S, Cheng A, Kassotis CD, Balise V, Nagel SC (2014) Developmental and reproductive effects of chemicals associated with unconventional oil and natural gas operations. Rev Environ Health 29:307–318. https://doi.org/10.1515/reveh-2014-0057

Weinbach EC (1954) The effect of pentachlorophenol on oxidative phosphorylation. J Biol Chem 210:545–550

Weinbach EC, Garbus J (1969) Mechanism of action of reagents that uncouple oxidative phosphorylation. Nature 221:1016–1018

Wiemels J, Smith MT (1999) Enhancement of myeloid cell growth by benzene metabolites via the production of active oxygen species. Free Radic Res 30:93–103

Williams DE, Lech JJ, Buhler DR (1998) Xenobiotics and xenoestrogens in fish: modulation of cytochrome P450 and carcinogenesis. Mutat Res 399:179–192

Winn LM (2003) Homologous recombination initiated by benzene metabolites: A potential role of oxidative stress. Toxicol Sci 72:143–149. https://doi.org/10.1093/toxsci/kfg008

Wood CM (2001) Toxic responses of the gill. In: Schlenk D, Benson WH (eds) Target organ toxicity in marine and freshwater teleosts: organs, new perspectives: toxicology and the environment. Taylor and Francis, Milton Park, pp 1–89

Wood CM (2012a) An introduction to metals in fish physiology and toxicology: basic principles. In: Wood CM, Farrell AP, Brauner CJ (eds) Homeostasis and toxicology of essential metals, fish physiology. Academic Press, Cambridge, pp 2–53. https://doi.org/10.1016/s1546-5098(11)31037-0

Wood CM (2012b) Silver. In: Wood CM, Farrell AP, Brauner CJ (eds) Homeostasis and toxicology of non-essential metals, fish physiology. Academic Press, Cambridge, pp 2–67. https://doi.org/10.1016/s1546-5098(11)31034-5

Wood CM, Al-Reasi HA, Smith DS (2011) The two faces of DOC. Aquat Toxicol 105:3–8. https://doi.org/10.1016/j.aquatox.2011.03.007

Xiao G, Pan C, Cai Y, Lin H, Fu Z (2001) Effect of benzene, toluene, xylene on the semen quality and the function of accessory gonad of exposed workers. Ind Health 39:206–210. https://doi.org/10.2486/indhealth.39.206

Yang L, Zha J, Wang Z (2017) Pentachlorophenol affected both reproductive and interrenal systems: in silico and in vivo evidence. Chemosphere 166:174–183. https://doi.org/10.1016/j.chemosphere.2016.09.099

Yu H (2003) Environmental carcinogenic polycyclic aromatic hydrocarbons: photochemistry and phototoxicity. J Environ Sci Health C Environ Carcinog Ecotoxicol Rev 20:149–183. https://doi.org/10.1081/GNC-120016203

Zangar RC, Davydov DR, Verma S (2004) Mechanisms that regulate production of reactive oxygen species by cytochrome P450. Toxicol Appl Pharmacol 199:316–331. https://doi.org/10.1016/j.taap.2004.01.018

Zha J, Wang Z, Schlenk D (2006) Effects of pentachlorophenol on the reproduction of Japanese medaka (Oryzias latipes). Chem Biol Interact 161:26–36. https://doi.org/10.1016/j.cbi.2006.02.010

Zhang M, Yin D, Kong F (2008) The changes of serum testosterone level and hepatic microsome enzyme activity of crucian carp (Carassius carassius) exposed to a sublethal dosage of pentachlorophenol. Ecotoxicol Environ Saf 71:384–389. https://doi.org/10.1016/j.ecoenv.2007.10.014

Zhuoheng C, Osadetz KG (2013) An assessment of tight oil resource potential in Upper Cretaceous Cardium Formation, Western Canada sedimentary basin. Pet Explor Dev 40:344–353

Ziemkiewicz PF, Quaranta JD, Darnell A, Wise R (2014) Exposure pathways related to shale gas development and procedures for reducing environmental and public risk. J Nat Gas Sci Eng 16:77–84. https://doi.org/10.1016/j.jngse.2013.11.003

Zou E (2005) Impacts of xenobiotics on crustacean molting: the invisible endocrine disruption. Integr Comp Biol 45:33–38. https://doi.org/10.1093/icb/45.1.33

Occurrence of Levonorgestrel in Water Systems and Its Effects on Aquatic Organisms: A Review

Ana Lourdes Oropesa and Laura Guimarães

Contents

Abstract Levonorgestrel is one of the active ingredients of oral contraceptives detected in the aquatic environment at concentrations in the order of ng/L. During the past decade, a wealth of new information about levonorgestrel has been

The original version of this chapter was revised. The correction to this chapter is available at https://doi.org/10.1007/398_2020_52

A. L. Oropesa (✉)
Toxicology Area, University of Extremadura, Caceres, Spain

BRILiC – Biotechnology Research Institute in Livestock and Cynegetic, University of Extremadura, Caceres, Spain
e-mail: aoropesa@unex.es

L. Guimarães
CIIMAR – Interdisciplinary Centre of Marine and Environmental Research, University of Porto, Matosinhos, Portugal
e-mail: lguimaraes@ciimar.up.pt

P. de Voogt (ed.), *Reviews of Environmental Contamination and Toxicology Volume 254*,
Reviews of Environmental Contamination and Toxicology 254,
https://doi.org/10.1007/398_2020_44

produced, with several studies having reported negative effects in the reproduction and growth of aquatic organisms after exposure to this emerging contaminant of concern. In the present study, the data about its levels in water and its effects on aquatic organisms were integrated and used to perform an updated preliminary aquatic risk assessment for levonorgestrel based on the guideline for Environmental Risk Assessment of Medicinal Products for Human Use from the European Medicines Agency. The aim was to investigate if this pharmaceutical has a risk for adverse effects on aquatic organisms (i.e. for organisms residing in surface water and groundwater). The results evidenced that levonorgestrel is likely to pose an environmental risk to surface water (risk quotient >1). Based on these results, a more refined risk assessment for this pharmaceutical is needed. Besides, our findings highlight the need for investigation under the adverse outcome pathway (AOP) framework, as well as for further studies about toxicological interactions between levonorgestrel and other synthetic steroids.

Keywords Aquatic risk assessment · Environmental concentrations · Groundwater · Sediments · Surface water · Synthetic progestin · Toxicity

1 Introduction

Over the last two decades, many studies have reported widespread occurrence of pharmaceutical compounds in the aquatic environment. Among them, emphasis was given to synthetic progestins, which are routinely used in human and veterinary medicine around the world. Due to incomplete removal by conventional methods applied in sewage treatment plants (STP), low concentrations of these highly bioactive compounds are continuously released into the aquatic environment (Yarahmadi et al. 2018; Narváez et al. 2019; Yu et al. 2019). Amid these is levonorgestrel (LNG), a 19-nortestosterone-derived second-generation progestin. Levonorgestrel is a gonane generated from nandrolone, which integrates the composition of oral contraceptives. It is commonly taken in association with an estrogen and is also one of the active ingredients of the emergency contraceptives. Relative to progesterone, LNG shows a higher potency, measured as the relative binding affinity for the progesterone receptor (Juchem and Pollow 1990; Phillips et al. 1990). Moreover, Zeilinger et al. (2009) verified the content in progestins of hormonal pharmaceuticals which can be a hundred times greater than their composition in estrogens, suggesting concentrations of progestins found in the aquatic environment should be in the same or higher order of magnitude than those of synthetic estrogens.

Regarding its aquatic environmental fate and behaviour, LNG is expected to adsorb to suspended solids and sediments as a function of its partition coefficient between organic carbon and water (Koc) (Table 1). Volatilization from water surface shows little probability based upon estimated vapour pressure and Henry's law constant. Its potential for bioconcentration in aquatic organisms is moderate (Table 1). On the other hand, the substance is not expected to undergo hydrolysis

Table 1 Physico-chemical properties of levonorgestrel (LNG)

CASRN	797-63-7
IUPAC	17β-Hydroxy-18-methyl-19-nor-17α-pregn-4-en-20-yn-3-on
Common name	Levonorgestrel
Molecular weight[a]	312.45 g/mol
Molecular formula	C_{21}-H_{28}-O_2
Chemical structure	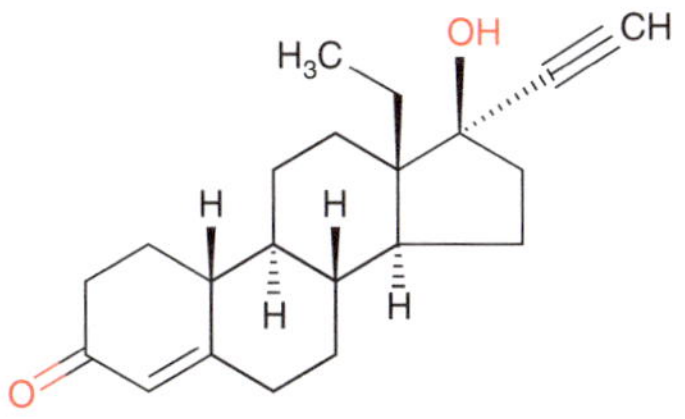
Solubility in water[b]	2.05 mg/L at 25°C
Solubility in organic solvents[b]	Ethanol (1 in 120), chloroform (1 in 15), diethyl ether (1 in 400) and dioxane
Koc[c]	5,353 L/kg
log Kow[b]	3.48
Vapour pressure/Henry's law constant[b]	3.92×10^{-10} mmHg at 25°C/7.7×10^{-10} atm-cu m/mol
Bioconcentration factor (BCF)[b]	92

[a]O'Neil et al. (2006)
[b]Value estimated using the Estimation Program Interface (EPI) Suite
[c]Value estimated using PcKocWin model, v 1.66

in the aquatic environment, although it may be susceptible to direct photolysis by sunlight (Lyman et al. 1990). Nevertheless, as a highly bioactive hormonal compound, exposure to LNG may cause long-term adverse effects in non-target organisms (Zeilinger et al. 2009; Runnalls et al. 2013; Hua et al. 2015; Cardoso et al. 2017; Thrupp et al. 2018). In face of this, and the most recent scientific literature, the ability of LNG to elicit unwanted effects in important pathways for growth, fertility and reproduction of aquatic organisms should be high, and previous risk evaluations may be underestimated. Based on the above and taking into account the high use of this progestin in some countries, the objectives of this study were: (1) to summarize data made available over the two last decades on the occurrence and detection of LNG in the aquatic environment, (2) to summarize research over this period with aquatic biota exposed to LNG and (3) to provide a preliminary aquatic risk assessment for this pharmaceutical.

2 Levonorgestrel Detection and Exposure Concentrations in the Aquatic Environment

Levonorgestrel has been detected in different countries around the world, in STP effluents, in water bodies and sediments of aquatic systems and in groundwater (Table 2). Its occurrence in abiotic environmental compartments has been investigated in Spain (Lopez de Alda et al. 2002; Petrovic et al. 2002), Belgium (Schmitt et al. 2011), France (Labadie and Budzinski 2005; Vulliet et al. 2007, 2008, 2011; Vulliet and Cren-Olivé 2011; Creusot et al. 2014), the Czech Republic (Matějíček and Kubáň 2007; Golovko et al. 2018; Šauer et al. 2018a, b), Poland (Kot-Wasik et al. 2016), Germany (Kuch and Ballschmiter 2000; Weizel et al. 2018), Sweden (Fick et al. 2010), Malaysia (Al-Odaini et al. 2010), China (Pu et al. 2008; Qiao et al. 2009; Sun et al. 2009; Liu et al. 2011; Shen et al. 2018; Yu et al. 2019), Australia (Scott et al. 2014; King et al. 2016), the United States (Chimchirian et al. 2007) and Canada (Viglino et al. 2008, 2011; Yarahmadi et al. 2018). Levels of LNG found in STP effluents range from 0.0081 to 30 ng/L, while those measured in surface waters range from 0.0075 to 38 ng/L (Table 2). Levels in groundwater were reported only for France as 11 ng/L (Vulliet et al. 2008) and 4 ng/L (Vulliet and Cren-Olivé 2011). Levels in sediments are available for Spain (Lopez de Alda et al. 2002), Belgium (Schmitt et al. 2011), China (Yu et al. 2019) and Canada (Viglino et al. 2011; Yarahmadi et al. 2018), with the highest concentrations found in the latest country (19 ng/g) (Table 2). A confrontation of the levels of LNG found in aquatic media with those found for synthetic estrogens usually associated with this progestin seems to support the hypothesis advanced by Zeilinger et al. (2009) that similar concentrations of these substances would be present in the aquatic environment. Another aspect of note is that among the 36 studies measuring LNG in STP effluents or surface waters, in about 40% of these LNG was below the limit of detection or quantification of the method (Table 2) and often found in low frequency in the samples analysed. Furthermore, these limits were very low (<10 ng/L) for most studies and countries, except for Poland (<78.95 ng/L) and Malaysia (<66 ng/L). This may occur because LNG consumption is generally low, and this is not a progestin widely prescribed in all countries. Contamination of the aquatic environment by LNG could thus represent an environmental problem only in some countries or regions of the world, mostly due to its high bioactivity.

3 Toxicity of Levonorgestrel to Aquatic Organisms

The information available about the ecotoxicological effects of LNG was obtained mainly on fish (Zeilinger et al. 2009; Pietsch et al. 2011; Runnalls et al. 2013, 2015; Svensson et al. 2013, 2014, 2016; Kroupova et al. 2014; Overturf et al. 2014; Hua et al. 2015; Liang et al. 2015a, 2018; Brockmeier et al. 2016; Cardoso et al. 2017, 2018, 2019; Hou et al. 2018; Thrupp et al. 2018) (Tables 3 and 4). However, reports on adverse effects upon LNG exposure can also be found for other taxa, such as cyanobacteria (Czarny et al. 2019a, b), algae (Czarny et al. 2019c), rotifers (Garcia-Garcia et al. 2014), molluscs (Contardo-Jara et al. 2011), crustaceans (Oropesa et al. 2010; Furuhagen et al. 2014) and amphibians (Kloas et al. 2009; Lorenz et al. 2011a, b; Kvarnryd et al. 2011; Säfholm et al. 2012, 2015, 2016). In total, 47 eco-toxicological studies determining effect concentrations for LNG were identified; 70% of these effect concentrations were derived from chronic toxicity assays. In the

Table 2 Maximum concentration of levonorgestrel (LNG) measured in the aquatic compartment (including sediments)

	Aquatic compartment	Place	Reference
Effluents of sewage treatment plants (ng/L)	1	Germany	Kuch and Ballschmiter (2000)
	4	Spain	Petrovic et al. (2002)
	<7.2[a]	France	Labadie and Budzinski (2005)
	<10[a]	United States	Chimchirian et al. (2007)
	17.9	France	Vulliet et al. (2007)
	1.3	China	Pu et al. (2008)
	30	Canada	Viglino et al. (2008)
	0.0081	China	Qiao et al. (2009)
	<66[a]	Malaysia	Al-Odaini et al. (2010)
	1	Sweden	Fick et al. (2010)
	6.7	China	Liu et al. (2011)
	10	France	Vulliet et al. (2011)
	<5[a]	Australia	Scott et al. (2014)
	<2.5[a]	Australia	King et al. (2016)
	<78.95[a]	Poland	Kot-Wasik et al. (2016)
	0.5	Czech Republic	Golovko et al. (2018)
	<0.3[a]	Czech Republic	Šauer et al. (2018a)
	<1[a]	Czech Republic	Šauer et al. (2018b)
	<0.04[a]	China	Shen et al. (2018)
	20.1	Canada	Yarahmadi et al. (2018)
	4	China	Yu et al. (2019)
Surface water (ng/L)	<5.9[a]	France	Labadie and Budzinski (2005)
	1	Czech Republic	Matějíček and Kubáň (2007)
	<1.84[a]	Spain	Kuster et al. (2008)
	7	France	Vulliet et al. (2008)
	0.0075	China	Qiao et al. (2009)
	<0.8[a]	China	Sun et al. (2009)
	38	Malaysia	Al-Odaini et al. (2010)
	22.2	China	Liu et al. (2011)
	3.6	France	Vulliet and Cren-Olivé (2011)
	33	France	Creusot et al. (2014)
	0.1	Czech Republic	Golovko et al. (2018)

(continued)

Table 2 (continued)

	Aquatic compartment	Place	Reference
	<0.6[a]	Czech Republic	Šauer et al. (2018a)
	<1.4[a]	Czech Republic	Šauer et al. (2018b)
	<0.03[a]	China	Shen et al. (2018)
	0.7	Germany	Weizel et al. (2018)
	<6[a]	Canada	Yarahmadi et al. (2018)
Groundwater (ng/L)	11	France	Vulliet et al. (2008)
	4	France	Vulliet and Cren-Olivé (2011)
Sediment (ng/g)	2.2	Spain	Lopez de Alda et al. (2002)
	<0.6[a]	Belgium	Schmitt et al. (2011)
	19	Canada	Viglino et al. (2011)
	<7.1[a]	Canada	Yarahmadi et al. (2018)
	0.1	China	Yu et al. (2019)

[a]Below the limit of detection or quantification of the method

following sections, we present the data obtained from the scientific literature, and our assessments, about the ecotoxicity of LNG towards aquatic organisms from different trophic levels.

3.1 Toxicity of Levonorgestrel to Cyanobacteria

Studies determining the toxicity of LNG to cyanobacteria are very recent, and only two were available in the literature (Czarny et al. 2019a, b). These studies investigated chronic effects of LNG and other hormones, single and in combination, on the growth of *Microcystis aeruginosa* and *Anabaena variabilis* (Table 4). The effect concentrations determined (median effective concentration, EC_{50}, and no observed effect concentration, NOEC) were rather high in the mg/L range, and LNG as found to be one of the less toxic hormones tested, which included 17-α-ethinylestradiol, progesterone, 17ß-estradiol, 5-pregnen-3β-ol-20-one, testosterone, estrone and estriol (Czarny et al. 2019a, b). As expected, hormone mixtures were reported to be more toxic than the single hormones.

Table 3 Acute and subacute toxicity data for levonorgestrel (LNG) in aquatic organisms

Taxonomic group	Species	Available data (μg/L)	Duration of the test	Effects	Reference
Crustaceans	*Daphnia magna* (Clone A6)	$LC_{50} = 1,660$	96 h	Mortality	Oropesa et al. (2010)
	Daphnia magna (Clone Klon 5)	$LC_{50} > 1,700$	48 h	Mortality	Furuhagen et al. (2014)
Molluscs	*Dreissena polymorpha*	NOEC < 3.12	7 days	Increased mRNA expression of metallothionein (gill) and catalase (digestive gland) Decreased mRNA expression of aryl hydrocarbon receptor (gills) and transmembrane P-glycoprotein (digestive gland)	Contardo-Jara et al. (2011)
Fish	*Cyprinus carpio*	NOEC ≤ 312	96 h (in vitro)	Innate immunity of fish through gene expression and measurement of nitric oxide production	Pietsch et al. (2009)
	Cyprinus carpio	NOEC < 1[a]	One injection	Immunosuppressive effects (decrease of nitric oxide production and increase of arginase activity)	Pietsch et al. (2011)
	Danio rerio	NOEC <0.002	144 h post fertilization	Stimulation of expression of genes related with the progesterone receptor in larvae	Zucchi et al. (2012)
	Cyprinus carpio	NOEC < $16*10^4$	10 min (in vitro)	Inhibition of CYP11ß activity	Fernandes et al. (2014)
	Danio rerio	NOEC <0.005	144 h post fertilization	Induction of genes related to the hypothalamic-pituitary-thyroid axis in larvae	Liang et al. (2015b)
	Gambusia holbrooki	NOEC < 0.01	8 days	Anal fin elongation in females	Frankel et al. (2016)
	Gambusia holbrooki	NOEC < 0.3	48 h	Up- and downregulation of several genes related to different biological processes (embryo development, immune response, lipid biosynthesis and transport)	Brockmeier et al. (2016)
	Pimephales promelas	NOEC < 0.01	14 days	Fin elongation and disruption of early-stage oogenesis in females	Frankel et al. (2017)
	Danio rerio	NOEC < 0.007	14 days	Increased frequency of mature sperm in males Alterations of expression in genes involved in the Notch signalling pathway in gonads	Shi et al. (2017)

(continued)

Table 3 (continued)

Taxonomic group	Species	Available data (μg/L)	Duration of the test	Effects	Reference
	Pimephales promelas	NOEC < 0.01	14 days	Decreased sperm motion characteristics	Frankel et al. (2018)
	Danio rerio	NOEC < 312	48 h	Induction of genes related to the circadian rhythm network in larvae	Zhao et al. (2018)
Amphibians	*Xenopus laevis*	NOEC < 0.0312	96 h	Increased sexual aroused state	Hoffmann and Kloas (2012)
	Xenopus laevis	NOEC < 3.1	72 h	Induction of gene expression in the thyroid gland	Lorenz et al. (2016)

[a]1 mg/kg b.w

Table 4 Chronic toxicity data for levonorgestrel (LNG) in aquatic organisms

Taxonomic group	Species	Available data (μg/L)	Duration of the test	Effects	Reference
Cyanobacteria	*Microcystis aeruginosa*	$EC_{50} = 3.54 \times 10^5$	14 days	Biomass concentration[a]	Czarny et al. (2019a)
	Anabaena variabilis	$NOEC < 0.1 \times 10^3$	14 days	Biomass concentration[a]	Czarny et al. (2019b)
Algae	*Chlorella variabilis*	$EC_{50} = 3.65 \times 10^5$	14 days	Biomass concentration[a]	Czarny et al. (2019c)
	Scenedesmus armatus	$EC_{50} = 3.07 \times 10^5$	14 days	Biomass concentration[a]	Czarny et al. (2019c)
Rotifers	*Anuraeopsis fissa*	NOEC < 31.25	21 days	Increase in population growth[a] Increased average lifespan[a] Increased net reproductive rate[a]	Garcia-Garcia et al. (2014)
	Brachionus calyciflorus	NOEC < 250	21 days	Increase in population growth and reproductive rates[a]	Garcia-Garcia et al. (2014)
Crustaceans	*Daphnia magna* (Clone A6)	NOEC = 156	21 days	Decrease in growth and reproductive output[a]	Oropesa et al. (2010)
	Daphnia magna (Clone Klon 5)	LOEC > 1,020	21 days	Decrease in reproduction[a] Decrease in feeding rate Increase in RNA/DNA content Inhibition of mRNA expression of vitellogenin and cuticle protein 12	Furuhagen et al. (2014)
Fish	*Pimephales promelas*	NOEC < 0.0008	21 days	Inhibition of reproduction[a]	Zeilinger et al. (2009)
	Pimephales promelas	NOEC < 0.1	21 days	Cessation of egg laying[a]	Runnalls et al. (2013)
	Gasterosteus aculeatus	NOEC < 0.0055	21 days	Induction of spiggin transcription in females	Svensson et al. (2013)
	Rutilus rutilus	NOEC < 0.03	28 days	Decrease in plasma 11-ketotestosterone concentration Increased percentage of spermatogonia B in males	Kroupova et al. (2014)

(continued)

Table 4 (continued)

Taxonomic group	Species	Available data (μg/L)	Duration of the test	Effects	Reference
	Pimephales promelas	NOEC $\leq$ 0.016	28 days post-hatch	Changes in mRNA transcript profiles in steroid receptors, steroidogenic enzymes and gonadotropins in larvae	Overturf et al. (2014)
	Gasterosteus aculeatus	NOEC < 0.0065	45 days	Induction of spiggin transcription in males	Svensson et al. (2014)
	Danio rerio	NOEC < 0.01	142 days post fertilization	Larvae masculinization[a]	Hua et al. (2015)
	Danio rerio	NOEC < 0.034	40 days post fertilization	Larvae masculinization[a]; induction and inhibition of genes related to sex differentiation and synthesis of sex hormones	Liang et al. (2015a)
	Pimephales promelas	NOEC < 0.0005	21 days	Female masculinization[a]	Runnalls et al. (2015)
	Gambusia holbrooki	NOEC < 0.1	21 days	Anal fin elongation in females[a]	Brockmeier et al. (2016)
	Danio rerio	NOEC < 0.005	50 days post fertilization	Induction of male phenotype[a]	Svensson et al. (2016)
	Danio rerio	NOEC < 0.01	21 days	Decrease in fecundity[a]	Cardoso et al. (2017)
	Danio rerio	NOEC < 0.01	21 days	Decline in the maturation stage of ovaries[a]	Cardoso et al. (2018)
	Gambusia affinis	NOEC < 0.0036	42 days	Increased frequency of atretic follicles in females[a]	Hou et al. (2018)
	Danio rerio	NOEC < 0.006	21 days	Upregulation of genes related with FSH and LH	Liang et al. (2018)
	Pimephales promelas	$EC_{10} = 0.00028$	21 days	Inhibition of egg production[a]	Thrupp et al. (2018)
	Danio rerio	NOEC < 0.01	21 days	Increase in hepatocyte size Increase in the glycogen content and catalase activity Decrease in vitellogenin content	Cardoso et al. (2019)

Amphibians	*Xenopus laevis*	NOEC < 156	80 days	Disruption of the gonadal structure[a] Delay in the metamorphosis process[a]	Kloas et al. (2009)
	Xenopus laevis	NOEC <0.31	80 days	Increase in the levels of prolactin and thyroid-stimulating hormones	Lorenz et al. (2011a)
	Xenopus laevis	NOEC < 0.31	80 days	Suppression of gonadotropin expression	Lorenz et al. (2011b)
	Xenopus tropicalis	NOEC < 0.17	88 days	Underdeveloped oviducts	Kvarnryd et al. (2011)
	Xenopus tropicalis	NOEC < 0.0013	28 days	Inhibition of vitellogenesis[a]	Säfholm et al. (2012)
	Xenopus tropicalis	NOEC < 0.03	49 days	Increase in mRNA levels of the androgen receptor in females	Säfholm et al. (2015)
	Xenopus tropicalis	NOEC < 0.003	49 days	Increase in expression of anti-Müllerian hormone and intracellular progesterone receptor in males	Säfholm et al. (2016)

[a]Endpoints used for estimating the PNEC (Technical Guidance for Deriving Environmental Quality Standards – Guidance Document No. 27 of the European Commission 2011)

3.2 Toxicity of Levonorgestrel to Algae

Likewise for cyanobacteria, investigation of the toxicity of LNG to algae is very recent (Czarny et al. 2019c) (Table 4). Here the authors also evaluated the chronic toxicity of several steroid hormones, such as progestagens (LNG, progesterone and 5-pregnen-3b-ol-20-one), estrogens (17-α-ethinylestradiol, estrone, 17ß-estradiol and estriol) and an androgen (testosterone) on the growth of *Chlorella vulgaris* and *Scenedesmus armatus*. The results were similar to those obtained for the cyanobacteria. Levonorgestrel was among the least toxic for the algae, with EC_{50} values within the mg/L range. Higher toxicity of the mixtures of hormones than the single compounds was also observed.

3.3 Toxicity of Levonorgestrel to Zooplankton

Acute and subacute toxicity to zooplanktonic organisms was evaluated using two different clones of *D. magna* (Table 3), while chronic toxicity was evaluated in rotifers and daphnids (Table 4). For rotifers, evaluation of chronic effects of LNG on the life table demography (endpoints: survivorship, fecundity, average lifespan, gross reproductive rate, net reproductive rate and generation time) and population growth rate of *Anuraeopsis fissa* and *Brachionus calyciflorus* led to the estimation of NOEC values markedly lower than those reported for cyanobacteria and algae (Garcia-Garcia et al. 2014) (Table 4). Interestingly, the study revealed that 31.25 and 62.5 μg/L LNG produced a significant increase in the population growth, average lifespan and net reproductive rate of *A. fissa*. For *B. calyciflorus* the reproductive rate and the rate of population increase per day of were also stimulated at 250 and 500 μg/L, the highest concentrations tested (Table 4).

Concerning the cladoceran species, acute and chronic toxicity were investigated using the *D. magna* clone A6 (Oropesa et al. 2010) and Klon 5 (Furuhagen et al. 2014). Assessments with clone A6 were done according to standard methods (OECD Guideline 202, 2004; OECD Guideline 211, 2008), using ASTM test medium. The acute toxicity test revealed that exposure to LNG (concentrations: 0.08, 0.16, 0.31, 0.63, 1.25 and 2.5 mg/L) for 48 h resulted in 30% immobilization at 2.5 mg/L. Employing the probit analysis (Finney 1971), the 96-h EC_{50} of LNG for *D. magna* was estimated to be 1.66 mg/L with a 95% confidence interval of 1.14–3.10 mg/L (Oropesa et al. 2010) (Table 3). The chronic toxicity test showed alterations in growth and reproduction, relative to the controls (Table 4). All concentrations of LNG tested (10, 19, 39, 78, 156 and 313 μg/L) elicited some level of juvenile mortality. From a reproductive point of view, a significant increase of age of physiological maturation was observed at 19 μg/L, together with a significant decrease in offspring production, compared to the control group. Furthermore, a significant decrease (−17%) in fecundity was detected at the highest concentration tested, relative to the control. Also, there was a significant interaction effect between treatment and the growth period (factorial ANOVA), indicating that the treatment had different effects on the two growth periods. In particular, growth during the juvenile period was significantly decreased in females exposed to 10 and 39 μg/L LNG, compared to growth recorded in the control group. Recovery from this reduction seemed to have occurred during the reproductive period for most tested concentrations, with exposed daphnids exhibiting a growth increase significantly higher than that of controls during this phase. Nevertheless, daphnids exposed to 313 μg/L still exhibited decreased growth at the end of the assay, compared to the controls (Oropesa et al. 2010). Overall, the findings of this study evidenced that both reproduction and growth of clone A6 were affected after exposure to concentrations of LNG between 10 and 313 μg/L. The NOEC and LOEC (lowest observed effect concentration) values for reproduction (fecundity) and total growth were estimated to be 156 and 313 μg/L LNG, respectively. Based on the geometric mean of these parameters (European Commission 2003), a MATC (Maximum Acceptable Toxicant Concentration) of 234.58 μg/L LNG was calculated. The acute and chronic toxicity of LNG for *D. magna* (Klon 5) were assessed according to the same OECD guidelines and using M7 test medium (Furuhagen et al. 2014). Acute toxicity tests were conducted with LNG concentrations ranging from 0.11 to 1.7 mg/L. Reproduction tests were conducted with LNG concentrations ranging from 0.013 to 1.02 mg/L. However, the authors detected no acute or chronic effects of LNG within the range of concentrations tested, including for daphnid development and feeding rate (Furuhagen et al. 2014) (Tables 3 and 4). Similarly, they found no differences relative to the control group for RNA/DNA measurements and mRNA expression of vitellogenin and cuticle protein 12 in daphnids exposed to 1.02 mg/L (Table 4). The different results obtained by Oropesa et al. (2010) and Furuhagen et al. (2014) may be attributed to differences in sensitivity to LNG of the two clones tested.

3.4 *Toxicity of Levonorgestrel to Molluscs*

Contardo-Jara et al. (2011) examined the bioaccumulation and effects of exposure to LNG for 7 days in a flow-through system on the bivalve *Dreissena polymorpha* (Table 3). The authors reported 100-fold bioconcentration factors within 4 days of exposure to 0.312 μg/L LNG and above 200 at the end of the test. Besides, 3.12 and 6.24 μg/L LNG caused protein damage in the gills and increased levels of superoxide dismutase in the digestive gland. The authors also found increased mRNA expression of several genes related to the toxicological response at the end of the exposure period, compared to controls (Table 3).

3.5 Toxicity of Levonorgestrel to Fish

Various fish species were employed to assess the acute, subacute and chronic toxicity of LNG (Tables 3 and 4): *Pimephales promelas*, *Oncorhynchus mykiss*, *Danio rerio*, *Cyprinus carpio*, *Gambusia holbrooki*, *Pimephales promelas*, *Gasterosteus aculeatus*, *Rutilus rutilus*, and *Gambusia affinis*. Regarding acute and subacute toxicity, besides mortality, effects have mostly been observed in reproductive organs and the expression of genes involved in the endocrine control of reproduction. One of the earliest works was carried out by Pietsch et al. (2009) to investigate effects of LNG on intracellular pathways regulating mRNA expression of genes involved in inflammatory. The experiments were conducted in vitro with leukocytes obtained from the head and trunk kidney of fish. Immunosuppression was gauged through gene expression and nitric oxide production. For LNG no alterations were found for nitric oxide production and iNOS mRNA expression in cells exposed to 3.12 and 312 µg/L LNG for 96 h. Subsequently, one study reported on the exposure to undiluted STP effluents for 14 days (Fick et al. 2010). The authors showed LNG accumulation in the plasma of the rainbow trout (*O. mykiss*) at concentrations in the range 8.5–12 ng/mL following the exposure. Another work by Pietsch et al. (2011) continued to investigate the immunological effects of LNG (Table 3). The authors injected carp (*C. carpio*) with a single dose of LNG, among other hormones tested, and assessed effects on leucocytes isolated from their kidneys up to 21 days after the injection. A dose of 1 mg/kg b.w. LNG was found to elicit immunosuppressive effects by decreasing the production of nitric oxide (decreased) and increasing the arginase activity of isolated leucocytes (Pietsch et al. 2011). Among the various works with zebrafish, Zucchi et al. (2012) reported on the exposure of embryos to 2, 20 and 200 ng/L LNG for 144 hpf (hours post fertilization). At the end of the experiments, larvae exposed to 2 ng/L LNG exhibited enhanced expression of the gene related to the progesterone receptor (Zucchi et al. 2012). Conversely, larvae exposed to 20 ng/L showed downregulation of the gene associated with the androgen receptor. These altered expression patterns could affect embryonic development (Zucchi et al. 2012). Another study investigated LNG effects on carp steroidogenic enzymes, in vitro (Fernandes et al. 2014) (Table 3). The results showed LNG could significantly inhibit the gonadal enzyme CYP11ß in this species, at a concentration of 160 mg/L, an effect able to interfere with the biosynthesis of androgens. Other gene expression investigations detected strong induction of genes related to the hypothalamic-pituitary-thyroid axis in zebrafish larvae exposed to 5, 50 and 100 ng/L LNG for 144 hpf (Liang et al. 2015b) (Table 3). Studies in the mosquitofish were done by Frankel et al. (2016) and Brockmeier et al. (2016). The former authors reported masculinization, evidenced by the elongation of the anal fin rays, in females exposed to 10 and 100 ng/L for 8 days. They also observed changes in the reproductive behaviour of males exposed to their highest test concentration. The study of Brockmeier et al. (2016) used a short exposure (48 h) to 300 ng/L LNG to investigate changes in gene expression using a microarray. They observed up- and downregulation of several genes related to different biological processes, such as embryo development, immune response, lipid biosynthesis and transport. Another study by Frankel et al. (2017) also found anal fin elongation and disruption of the early-stage oogenesis in females exposed to 10 and 100 ng/L LNG for 14 days, but this time in fathead minnows (*P. promelas*). Furthermore, exposure to 100 ng/L LNG also caused a decrease in the gonadosomatic index and blood plasma vitellogenin levels in females (Frankel et al. 2017). Still working with zebrafish, Shi et al. (2017) evaluated the histological and transcriptional alterations in gonads of adults after exposure to 6.7, 83 and 912 ng/L LNG for 14 days. This study revealed an improvement in the percentage of male mature sperms and reduction in that of immature sperms at all concentrations, as well as the augmented frequency of female mature vitellogenic oocyte, but also of atretic follicles, at 912 ng/L. These changes were accompanied by the activation of the Notch signalling pathway in gonads (Table 3). Later on, Frankel et al. (2018) reported sperm mobility was also affected in males of fathead minnows exposed to 100 ng/L for 14 days. Finally, in a very recent work, the circadian rhythm network was found to be altered in zebrafish larvae exposed to 312 ng/L LNG from 120 to 164 hpf, as indicated by the transcriptional responses of core clock genes (Zhao et al. 2018).

As to chronic toxicity, Zeilinger et al. (2009) investigated the effects of LNG on adult fathead minnows (Table 4). The authors observed alterations on the reproductive function of fish exposed for 21 days to 0.8, 3.3 and 29.6 ng/L LNG. All tested concentrations led to inhibition of reproduction manifested as a decrease in female egg deposition or changes in the male reproductive behaviour. A previous range-finding test had revealed that females exposed to 100 ng/L LNG developed male secondary sexual characteristics such as darker pigmentation and emergence of nuptial tubercles (Zeilinger et al. 2009). Subsequently, investigations of Runnalls et al. (2013) found strong arrest of spawning (about 90% relative to the pre-exposure period) in adult fathead minnows exposed to 100 ng/L LNG for 21 days. Another study by Svensson et al. (2013) showed that exposure (21d) to LNG concentrations >40 ng/L may cause induction of spiggin transcription, kidney hypertrophy, decrease in liver *vitellogenin* transcription and *CYP1A* expression in female three-spined sticklebacks (*G. aculeatus*). The same authors subsequently reported that the transition to the non-breeding state (i.e. regression of secondary sex characteristics, cessation of spiggin production in the kidney and resumption of spermatogenesis in the testes) was inhibited by exposure for 45 days to an LNG concentration (6.5 ng/L) within the levels found in natural waters, with possible disruption of the seasonal breeding cycle (Svensson et al. 2014). Observations in both genders of roach (*R. rutilus*) showed deregulation in pituitary gonadotropin expression of luteinizing hormone (LH) and follicle-stimulating hormone (FSH), as well as decrease in the plasma sex steroid concentrations (11-ketotestosterone and estradiol), as main effects elicited by exposure to 31 and 3,124 ng/L LNG for 28 days. These alterations could impact negatively the gonadal maturation and spawning, and they were also accompanied by an increase in the percentage of spermatogonia B in males, compared to controls (Kroupova et al. 2014) (Table 4). Reduced survival and decreased growth were observed in fathead minnow larvae exposed to 462 ng/L and 86.9 ng/L

LNG for 28 days post-hatching, respectively (Overturf et al. 2014). Further, transcriptional mRNA downregulation of *3 ß-HSD*, *20 ß-HSD*, *CYP17*, *CYP19A*, *AR*, *ERα* and *FSH* genes was observed at 86.9 and 16.3 ng/L LNG. These authors also carried out an ex vivo steroidogenesis assay in adult females exposed to 100 ng/L LNG for 7 days. They observed a decrease in ovarian production of steroid hormones (pregnenolone, 17α,20ß-dihydroxy-4-pregnen-3-one (17,20-DHP), testosterone and 11-ketotestosterone). Again using zebrafish embryos, exposed for 142 dpf, Hua et al. (2015) studied the influence of LNG concentrations ranging from 1 to 100 ng/L on sex differentiation. They showed that all the evaluated concentrations caused masculinization during gonadal differentiation. Likewise, Liang et al. (2015a) observed masculinization, associated with altered induction and inhibition of genes related to sex differentiation and synthesis of sex hormones, in larvae exposed to 34 and 77 ng/L LNG for 40 dpf (Table 4). Masculinization was also detected in female fathead minnows exposed to 0.5, 5 and 25 ng/L for 21 days as indicated by the development of male secondary sexual characteristics (i.e. dorsal fin spots and fatpad at 0.5 ng/L; presence of tubercles, dorsal fin spots and fatpad at 5 and 25 ng/L). Reduction in egg production was additionally observed, in particular at the highest concentrations (about −85%). Very recently, several other studies reported the induction of male secondary sexual characteristics and/or reduction in egg production (Table 4). The work of Brockmeier et al. (2016) found anal fin elongation and abnormal embryo development in eastern mosquitofish (*G. holbrooki*) females exposed to 100 ng/L LNG for 21 days. Svensson et al. (2016) showed induction of the male phenotype (100% males) and precocious puberty in juvenile zebrafish exposed to 5.5, 79 and 834 ng/L LNG for 50 dpf. In experiments conducted to investigate the combined effects of LNG (10 and 1,000 ng/L for 21 days) and temperature increase in zebrafish, Cardoso and colleagues observed a decrease in the fecundity (Cardoso et al. 2017), impaired maturation stage of the ovaries (Cardoso et al. 2018) and immunohistochemical alterations in the liver (Cardoso et al. 2019), including increased size of hepatocytes and their glycogen content, as well as decreased vitellogenin content (Table 4). Apart from masculinization following exposure to 35.8 and 368.0 ng/L LNG for 42 days (evidenced by alterations in the anal fin and spermatogenesis occurring in the ovaries), females of mosquitofish (*G. affinis*) also exhibited an increase in the number of atretic follicles at a concentration as low as 3.6 ng/L (Hou et al. 2018). Interestingly, in females and males showing no signs of altered reproduction or gonad histopathology after exposure to 6, 29 or 69 ng/L LNG for 21 days, Liang et al. (2018) observed altered mRNA expression of several genes belonging to the hypothalamic-pituitary-gonadal (HPG) axis, mainly upregulation in the female brain and ovary and downregulation in the male testes. In a very recent study aimed at investigating and better predicting fish responses to mixtures of steroids, Thrupp et al. (2018) obtained concentration-response data on egg production (21d exposure) and were able to estimate a EC_{10} of 0.28 ng/L (Table 4).

In short, concentrations of LNG within the range found in natural surface waters were able to cause negative impacts in the reproductive function of fish, with consequences at the population level, including disruption of the seasonal breeding cycle in males, masculinization of larvae and females, disruption of oogenesis and decreased egg production. The data also pointed *P. promelas* as the most sensitive fish species.

3.6 Toxicity of Levonorgestrel to Amphibians

Adverse effects of LNG were also investigated in vertebrates of the genus *Xenopus* (Kloas et al. 2009; Lorenz et al. 2011a, b, 2016; Kvarnryd et al. 2011; Hoffmann and Kloas 2012; Säfholm et al. 2012, 2015, 2016). Exposure concentrations ranged between 1.3 ng/L and 3,124 × 10^3 ng/L, and most of the works were set to investigate chronic toxicity (Table 3 and 4). Evaluation of acute toxicity was done by Hoffmann and Kloas (2012) and Lorenz et al. (2016). The study published by Hoffmann and Kloas (2012) found that short-term exposure to environmental concentrations of LNG can induce the sexually aroused state in *X. laevis*. The work of Lorenz et al. (2016) reported induction of gene expression in the thyroid gland also of *X. laevis*, compared to controls (Table 4).

As to chronic toxicity, LNG was found to elicit negative effects on gonads and in hormonally controlled processes related to reproduction, development and metamorphosis. These included histopathological findings in both genders such as important morphologic disruption of gonad structure in *X. laevis* (Kloas et al. 2009) and the occurrence of immature oocytes, underdeveloped oviducts and female sterility in adult *X. tropicalis* frogs (Kvarnryd et al. 2011) (Table 4). Disturbances in female fertility, reduced egg deposition and suppression of gonadotropin expression, resulting in metamorphosis delay relative to controls, were described by Lorenz et al. (2011a, b) in *X. laevis*. These authors also reported increased levels of prolactin and thyroid-stimulating hormones in exposed frogs. Later on Lorenz and colleagues described the mechanisms related to the anti-metamorphic effects of LNG (Lorenz et al. 2016). Finally, Säfholm et al. (2012) indicated that environmental LNG concentrations could represent a serious threat to reproduction in *X. tropicalis* populations based on the alterations observed in the process of oogenesis and the detected inhibition of vitellogenesis. Further studies revealed that LNG causes induction of mRNA expression of the hepatic androgen receptor, and it may alter the development of the oviduct as a result of the perturbation of the anti-Müllerian hormone in females of this amphibian species (Säfholm et al. 2015, 2016).

Overall, the information reviewed herein showed that except for data obtained from zebrafish embryo assays lasting up to 144 hpf, the effect concentrations or NOEC values obtained for acute and subacute effects were at least two orders of magnitude higher than those reported for chronic LNG toxicity. In respect to chronic effects, the least sensitive taxa were the cyanobacteria and algae. The vertebrate taxa were the most sensitive to LNG, which can elicit serious disruption of reproductive organs and tissues reflecting in altered fertility. Based on the most restrictive endpoint value found for fathead minnows (*P. promelas*), fish represent the trophic level most sensitive to LNG in the aquatic compartment. Also, the compilation of ecotoxicological information provided in this manuscript allows to include this pharmaceutical into category chronic 1 of substances hazardous to the aquatic environment proposed by the Globally Harmonised System of Classification and Labelling of Chemicals (GHS 2007), according to the chronic NOEC for fish <0.1 mg/L.

4 Aquatic Risk Assessment for Levonorgestrel

To understand if LNG could have potential risk for the aquatic environment, a preliminary risk assessment of this synthetic progestin was carried out according to the European Medicines Agency (EMEA 2006) on Environmental Risk Assessment of Medicinal Products For Human Use. In accordance, as the estimated LNG Koc value is <10,000 L/kg (Koc = 5,353 L/kg) (Table 1), bank infiltration may likely constitute a path for groundwater contamination by surface waters (EMEA 2006). Hence, the initial risk assessment was performed for both surface water and groundwater. As stated in the EMEA (2006) phased methodology, this hormonal compound is suggested to enter directly the screening stage in the environmental risk assessment (ERA). This indication is due to its potential to cause adverse effects in the aquatic environment at low concentrations ($EC_{10} = 0.28$ ng/L), as established by Thrupp et al. (2018). The risk assessment should thus proceed in the following sequence: estimation of exposure or selection of measured environmental concentration (MEC), effect assessment and estimation of hazard quotients.

4.1 Measured Environmental Concentration (MEC) Selection

To obtain a more realistic ERA, the highest MEC values obtained from the literature compiled for LNG (Table 2) were used in the risk calculations, rather than the exposure estimation indicated in Technical Guidance Document on Risk Assessment: Part II (TGD) issued by the European Commission (2003). These concentrations correspond to 38 ng/L for surface water (Al-Odaini et al. 2010) and 11 ng/L for groundwater (Vulliet et al. 2008).

4.2 Effects Assessment

In this phase, the predicted no effect concentration (PNEC), i.e. the highest concentration which does not originate undesirable effects in organisms of a given environmental compartment, was estimated. The PNEC estimations were conducted for

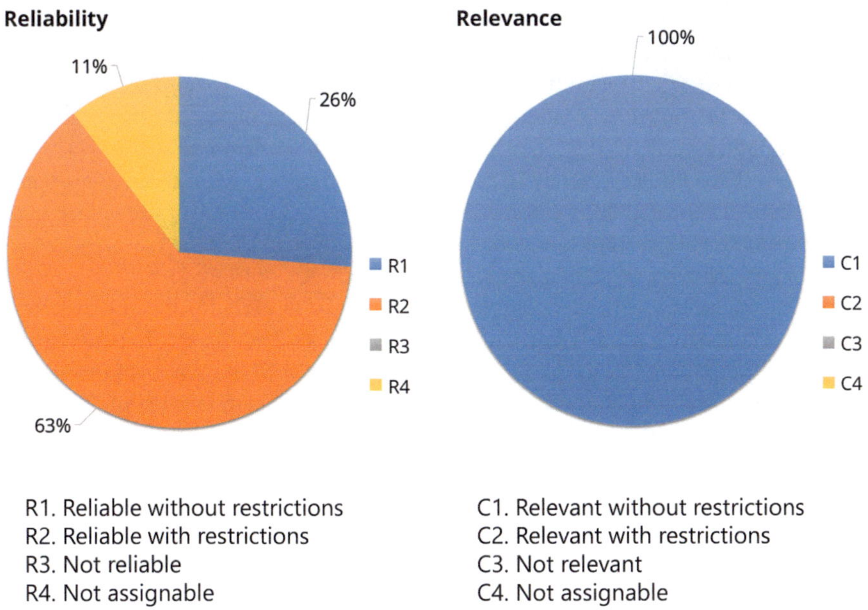

Fig. 1 Evaluation of reliability and relevance of the ecotoxicological studies used for PNEC derivation. The validation was done using the method described by Moermond et al. (2016)

surface water ($PNEC_{SW}$) and groundwater ($PNEC_{GW}$) in agreement to the TGD (European Commission 2003) and the EMEA Guideline (EMEA 2006), from the chronic toxicity data reviewed for aquatic organisms, which is presented in Table 4. They further took into consideration the Technical Guidance for Deriving Environmental Quality Standards (TGD-EQS) of the European Commission (2011), which states that only those endpoints with relevance at the population level should be taken into account to estimate the PNEC values. The Criteria for Reporting and Evaluating Ecotoxicity Data (CRED) were further used to review and evaluate reliability and relevance of the studies used in the derivation of the PNECs, as recommended by Moermond et al. (2016). Review of the studies according to CRED for reliability led to the classification of 89% of the studies either as reliable without restriction (R1) or reliable with restrictions (R2, 63%, Fig. 1). No study was considered as not reliable. Classification as R2 or R4 was due to lack of information reported. As to biological relevance, all studies provided sound data about their relevance without restrictions.

Regarding PNEC derivation for surface water, the lowest available EC_{10} (0.00028 μg/L), identified for *P. promelas* (Thrupp et al. 2018), was divided by an assessment factor (AF), according to the recommendations of EMEA (2006) and the TDG (European Commission 2003). In terms of ERA, the application of an AF covers some uncertainties resulting from intra- and inter-laboratory variation of toxicity data, intra- and interspecies variations, short-term to long-term toxicity extrapolation or laboratory data to field extrapolation (European Commission

2003). In the present case, given that long-term data is available for three species representing three different trophic levels (algae as producers, daphnia as primary consumers and fish as secondary consumers), an AF of 10 could be applied. However, as previously considered for a risk assessment of the anti-diabetic metformin (Godoy et al. 2018), the TGD-EQS (European Commission 2011) also establishes that a higher AF may be required when there is evidence that a substance may induce adverse effects via endocrine disruption. From the previous section, LNG was shown to disrupt fish gonad structure, larvae and female masculinization and a series of other effects indicative of interference with endocrine disruption pathways, including enhanced expression of genes related with the progesterone receptor (Zucchi et al. 2012); induction of genes related to the hypothalamic-pituitary-thyroid axis (Liang et al. 2015b); changes in mRNA transcript profiles in steroid receptors, steroidogenic enzymes and gonadotropins (Overturf et al. 2014); alterations in the expression of genes related to sex differentiation; and synthesis of sex hormones (Liang et al. 2015a). Hence, to better understand the potential risk, two $PNEC_{sw}$ values were calculated using an AF of either 10 or 50. The $PNEC_{SW}$ values obtained were 0.000028 μg/L (AF = 10) and 0.0000056 μg/L (AF = 50). For the estimation of the PNEC for groundwater, the EMEA guidance recommends applying a factor of 10 to the lowest NOEC resulting from *Daphnia* reproduction tests (EMEA 2006). However, given the effects observed on reproduction (Oropesa et al. 2010) and the ability of LNG to interfere with the mRNA expression of vitellogenin (Furuhagen et al. 2014), the same approach used for surface water was followed here. Hence, the lowest NOEC of 156 μg/L LNG (Oropesa et al. 2010; Table 3) was divided either by 10 or by 50, resulting in $PNEC_{GW}$ values of 15.6 μg/L and 3.2 μg/L, respectively.

4.3 Estimation of Risk Quotients

Once conducted the previous phases (Sects. 4.1 and 4.2), a quantitative characterization must be performed to identify the potential risk of LNG to surface and groundwater. Consequently, the selected measured environmental concentrations (MEC_{SW} and MEC_{GW}) were divided by the respective estimated PNEC values ($PNEC_{SW}$ and $PNEC_{GW}$) to establish the risk quotients (RQs) (European Commission 2003; EMEA 2006). Bearing in mind the worst-case scenario, and irrespective of whether the adverse effects of LNG are induced via endocrine disruption or not, an $RQ > 1$ was obtained for surface water, indicating potential environmental risk for its aquatic biota. An $RQ < 1$ was obtained for groundwater, meaning that risk should be negligible to organisms in this compartment. A previous study predicted that effluents from STP in their point of discharge to superficial waters had no risk to aquatic vertebrates based on the use of a prediction model (King et al. 2016). Nevertheless, LNG is a complex substance able to elicit hazardous reproductive effects at very low concentrations, as depicted by several works, so that our study reveals a potential risk to organisms that live in superficial water. Apart from using a different tool to estimate the aquatic risk assessment of this progestin, King et al. (2016) utilized predicted effluent concentrations of LNG in contrast to the measured surface water concentrations used in our study. This could also be the cause for the different results obtained. Adding to this, more recent studies have brought further information and effect concentrations for LNG. It is important to note that the ERA of a compound should be a revisable and updatable process based on new exposure and effect data derived from the literature. To further clarify and manage LNG risks to the environment, more studies are required to know the fate and behaviour of this compound in the aquatic compartment and its capacity to bioaccumulate in aquatic organisms and also to cause toxicological effects in other trophic levels, for example, *Chironomus*, where there are important information gaps. These could help to refine the aquatic risk assessment for this pharmaceutical.

In a real scenario, aquatic organisms are usually exposed to pharmaceutical mixtures that, from a toxicological point of view, could interact between them. This is especially relevant in the case of other progestins and estrogens which have been demonstrated to interact with LNG action on growth and/or reproduction. Regarding fish, Runnalls et al. (2015) reported additive effects on the fecundity of fathead minnow by a mixture of LNG and 17 α-ethinylestradiol (EE2). Testing the effects of mixtures in zebrafish, Hinfray et al. (2016) found that combined exposure to EE2 and LNG exerted additive estrogenic effects both in vitro and in vivo. And, Thrupp et al. (2018) observed a five-component mixture of the sex steroid hormones EE2, LNG, beclomethasone dipropionate, desogestrel and trenbolone, with different modes of action, could elicit additive inhibition of egg production in fathead minnow. Moreover, this additive effect was observed in mixtures in which each compound was present at a low concentration that on its own would not produce significant effects when tested alone (Hinfray et al. 2016). Also, Maasz et al. (2017) observed that adult roaches exposed to a mixture of progestogens (progesterone, drospirenone and LNG) in an environmentally relevant concentration showed a significant increase of DJ-1 chaperone proteins in the brain and liver tissues, compared to controls, with this parameter considered as a possible indirect biomarker of steroid contamination. On the other hand, antagonistic effects on sex ratio and hepatic mRNA expression of *vitellogenin* and steroid hormone receptors of mixtures of LNG and EE2 were observed in amphibians (Säfholm et al. 2015).

Overall, the work presented herein points out the need to perform additional studies and conduct a sound environmental risk assessment of LNG. In particular, future studies should be done within the adverse outcome pathway (AOP) framework. An AOP depicts a biological pathway linking an initiating event (i.e. molecular) to a cascade of ensuing events and the resulting adverse outcome (Ankley et al. 2010). It can be deranged by a toxicant chemical or another stressor, capturing the consecutive alterations occurring at various levels of biological organization that will ultimately lead to adverse effects of interest from a regulatory viewpoint (Perkins et al. 2019). As such, an AOP is a conceptual framework for depiction at a level relevant to risk assessment, providing solid ground to organize existing toxicological knowledge and describe the causal relationships between the succession of events and the outcome. Furthermore, it has been proven useful to integrate dose-response assessment with exposure assessment and can thus be used in the different stages of environmental risk assessment (from hazard and dose-response assessment to exposure assessment and risk characterization), to predict the adverse effects of chemicals and increasing its applicability and accuracy (Perkins et al. 2019).

5 Conclusions

This study provided data to refine the assessment factors to be applied in the calculations for the ecological risk of LNG. The results of an initial risk assessment performed according to the EMEA, and accounting for the most recent concentrations quantified in aquatic reservoirs and ecotoxicological data available for LNG, indicated the need to evaluate the environmental fate of this synthetic progestin and carry out a second-tier assessment for surface water.

Acknowledgments This work was supported by UIDB/04423/2020 and UIDP/04423/2020, through FCT (Portuguese Foundation for the Science and Technology), and by the REWATER project (FCT/UEFISCDI/FORMAS, WaterJPI/0008/2016) financed under the ERA-NET Cofund WaterWorks2015 (Water JPI).

Conflict of Interest On behalf of all authors, the corresponding author states that there is no conflict of interest.

References

Al-Odaini NA, Zakaria MP, Yaziz MI, Surif S (2010) Multi-residue analytical method for human pharmaceuticals and synthetic hormones in river water and sewage effluents by solid-phase extraction and liquid chromatography-tandem mass spectrometry. J Chromatogr A 1217:6791–6806. https://doi.org/10.1016/j.chroma.2010.08.033

Ankley GT, Bennett RS, Erickson RJ, Hoff DJ, Hornung MW, Johnson RD, Mount DR, Nichols JW, Russom CL, Schmieder PK, Serrano JA, Tietge JE, Villeneuve DL (2010) Adverse outcome pathways: A conceptual framework to support ecotoxicology research and risk assessment. Environ Toxicol Chem 29:730–741

Brockmeier EK, Scott PD, Denslow ND, Leusch FDL (2016) Transcriptomic and physiological changes in Eastern Mosquitofish (*Gambusia holbrooki*) after exposure to progestins and anti-progestagens. Aquat Toxicol 179:8–17. https://doi.org/10.1016/j.aquatox.2016.08.002

Cardoso PG, Rodrigues D, Madureira TV, Oliveira N, Rocha MJ, Rocha E (2017) Warming modulates the effects of the endocrine disruptor progestin levonorgestrel on the zebrafish fitness, ovary maturation kinetics and reproduction success. Environ Pollut 229:300–311. https://doi.org/10.1016/j.envpol.2017.05.090

Cardoso PG, Rodrigues D, Madureira TV, Rocha MJ, Rocha E (2018) Histopathological evaluation of combined impacts of the synthetic progestin levonorgestrel and temperature on the female zebrafish maturation using a semi-quantitative grading analysis – is it enough? Bull Environ Contam Toxicol 101:417–422. https://doi.org/10.1007/s00128-018-2436-z

Cardoso PG, Resende-de-Oliveira R, Madureira TV, Rocha MJ, Rocha E (2019) Combined effects of increased temperature and levonorgestrel exposure on zebrafish female liver, using stereology and immunohistochemistry against catalase, CYP1A, HSP90 and vitellogenin. Environ Pollut 252:1059–1067. https://doi.org/10.1016/j.envpol.2019.06.058

Chimchirian RF, Suri RP, Fu H (2007) Free synthetic and natural estrogen hormones in influent and effluent of three municipal wastewater treatment plants. Water Environ Res 79(9):969–974. https://doi.org/10.2175/106143007x175843

Contardo-Jara V, Lorenz C, Pflugmacher S, Nutzmann G, Kloas W, Wiegand C (2011) Molecular effects and bioaccumulation of levonorgestrel in the non-target organism *Dreissena polymorpha*. Environ Pollut 159:38–44. https://doi.org/10.1016/j.envpol.2010.09.028

Creusot N, Aït-Aïssa S, Tapie N, Pardon P, Brion F, Sanchez W, Thybaud E, Porcher JM, Budzinski H (2014) Identification of synthetic steroids in river water downstream from pharmaceutical manufacture discharges based on a bioanalytical approach and passive sampling. Environ Sci Technol 48(7):3649–3657. https://doi.org/10.1021/es405313r

Czarny K, Szczukocki D, Krawczyk B, Gadzala-Kopciuch R, Skrzypek S (2019a) Toxicity of single steroid hormones and their mixtures toward the cyanobacterium *Microcystis aeruginosa*. J Appl Phycol 31:3537–3544. https://doi.org/10.1007/s10811-019-01874-x

Czarny K, Szczukocki D, Krawczyk B, Skrzypek S, Miekos E, Gadzala-Kopciuch R (2019b) Inhibition of growth of *Anabaena variabilis* population by single and mixed steroid hormones. J Appl Phycol 31:389–398. https://doi.org/10.1007/s10811-018-1589-9

Czarny K, Szczukocki D, Krawczyk B, Skrzypek S, Zieliński M, Gadzała-Kopciuch R (2019c) Toxic effects of single animal hormones and their mixtures on the growth of *Chlorella vulgaris* and *Scenedesmus armatus*. Chemosphere 224:93–102. https://doi.org/10.1016/j.chemosphere.2019.02.072

European Commission (2003) Technical guidance document on risk assessment – Part II – environmental risk assessment. European Union, Brussels

European Commission (2011) Technical guidance for deriving environmental quality standards. Common Implementation Strategy for the Water Framework Directive (2000/60/EC). Guidance Document No. 27. European Union, Brussels

European Medicines Agency (EMEA) (2006) Guideline on the environmental risk assessment of medicinal products for human use. Committee for Medicinal Products for Human Use. EMEA/CHMP/SWP/4447/00 European Medicines Agency, London

Fernandes D, Pujol S, Aceña J, Perez S, Barceló D, Portea C (2014) The *in vitro* interference of synthetic progestogens with carp steroidogenic enzymes. Aquat Toxicol 155:314–321. https://doi.org/10.1016/j.aquatox.2014.07.014

Fick J, Lindberg RH, Parkkonen J, Arvidsson B, Tysklind M, Larsson DG (2010) Therapeutic levels of levonorgestrel detected in blood plasma of fish: results from screening rainbow trout exposed to treated sewage effluents. Environ Sci Technol 44(7):2661–2666. https://doi.org/10.1021/es903440m

Finney DJ (1971) Probit Analysis, 3rd edn. Cambridge University Press, Cambridge

Frankel TE, Meyer MT, Orlando EF (2016) Aqueous exposure to the progestin, levonorgestrel, alters anal fin development and reproductive behavior in the eastern mosquitofish (*Gambusia holbrooki*). Gen Comp Endocrinol 234:161–169. https://doi.org/10.1016/j.ygcen.2016.01.007

Frankel T, Yonkos L, Frankel J (2017) Environ Toxicol Chem 36(12):3299–3304. https://doi.org/10.1002/etc.3906

Frankel T, Yonkos L, Ampy F, Frankel J (2018) Exposure to levonorgestrel increases nest acquisition success and decreases sperm motility in the male Fathead Minnow (*Pimephales promelas*). Environ Toxicol Chem 37(4):1131–1137. https://doi.org/10.1002/etc.4054100

Furuhagen S, Fuchs A, Lundström Belleza E, Breitholtz M, Gorokhova E (2014) Are pharmaceuticals with evolutionary conserved molecular drug targets more potent to cause toxic effects in non-target organisms? PLoS One 9(8):e105028. https://doi.org/10.1371/journal.pone.0105028

Garcia-Garcia G, Sarma SSS, Nuñez-Orti AR, Nandini S (2014) Effects of the mixture of two endocrine disruptors (ethinylestradiol and levonorgestrel) on selected ecological endpoints of *Anuraeopsis fissa* and *Brachionus calyciflorus* (Rotifera). Int Rev Hydrobiol 99:1–7. https://doi.org/10.1002/iroh.20130172

Globally Harmonized System (GHS) of Classification and Labelling of Chemicals (2007) Part 4, Environmental hazards. Chapter 4.1: Hazardous to the aquatic environment. 4.1.2.: Classification criteria for substances. United Nations, Geneva, p 219

Godoy AA, Domingues I, Arsénia Nogueira AJ, Kummrow F (2018) Ecotoxicological effects, water quality standards and risk assessment for the anti-diabetic metformin. Environ Pollut 243:534–542. https://doi.org/10.1016/j.envpol.2018.09.031

Golovko O, Šauer P, Fedorova G, Kroupová HK, Grabic R (2018) Determination of progestogens in surface and waste water using SPE extraction and LC-APCI/APPI-HRPS. Sci Total Environ 621:1066–1073. https://doi.org/10.1016/j.scitotenv.2017.10.120

Hinfray N, Tebby C, Garoche C, Piccini B, Bourgine G, Aït-Aïssa S, Kah O, Pakdel F, Brion F (2016) Additive effects of levonorgestrel and ethinylestradiol on brain aromatase (cyp19a1b) in zebrafish specific *in vitro* and *in vivo* bioassays. Toxicol Appl Pharmacol 307:108–114. https://doi.org/10.1016/j.taap.2016.07.023

Hoffmann F, Kloas W (2012) The synthetic progestogen, Levonorgestrel, but not natural progesterone, affects male mate calling behavior of *Xenopus laevis*. Gen Comp Endocrinol 176:385–390. https://doi.org/10.1016/j.ygcen.2012.02.009

Hou LP, Chen H, Tian CE, Liang Y, Wu RR, Zhang XM, Fang XW, Zhang CP, Hu JJ, Song LY, Liang YQ, Schlenkd D, Xie L (2018) Alterations of secondary sex characteristics, reproductive histology and behaviors by norgestrel in the western mosquitofish (*Gambusia affinis*). Aquat Toxicol 198:224–230. https://doi.org/10.1016/j.aquatox.2018.03.014

Hua J, Han J, Guo Y, Zhou B (2015) The progestin levonorgestrel affects sex differentiation in zebrafish at environmentally relevant concentrations. Aquat Toxicol 166:1–9. https://doi.org/10.1016/j.aquatox.2015.06.013

Juchem M, Pollow K (1990) Binding of oral contraceptive progestins to serum proteins and cytoplasmic receptor. Am J Obstet Gynecol 163(6):2171–2182. https://doi.org/10.1016/0002-9378(90)90559-P

King OC, van de Merwe JP, McDonald JA, Leusch FDL (2016) Concentrations of levonorgestrel and ethinylestradiol in wastewater effluents: is the progestin also cause for concern? Environ Toxicol Chem 35(6):1378–1385. https://doi.org/10.1002/etc.3304

Kloas W, Urbatzka R, Opitz R, Würtz S, Behrends T, Hermelink B, Hofmann F, Jagnytsch O, Kroupova H, Lorenz C, Neumann N (2009) Endocrine disruption in aquatic vertebrates. Ann N Y Acad Sci 1163(1):187–200

Kot-Wasik A, Jakimska A, Śliwka-Kaszyńska M (2016) Occurrence and seasonal variations of 25 pharmaceutical residues in wastewater and drinking water treatment plants. Environ Monit Assess 188(12):661. https://doi.org/10.1007/s10661-016-5637-0

Kroupova HK, Trubiroha A, Lorenz C, Contardo-Jara V, Lutz I, Grabic R, Kocour R, Kloas W (2014) The progestin levonorgestrel disrupts gonadotropin expression and sex steroid levels in pubertal roach (*Rutilus rutilus*). Aquat Toxicol 154:154–162. https://doi.org/10.1016/j.aquatox.2014.05.008

Kuch HM, Ballschmiter K (2000) Determination of endogenous and exogenous estrogens in effluents from sewage treatment plants at the ng/L-level. Fresenius J Anal Chem 366:392–395

Kuster M, de Alda MJL, Hernando MD, Petrovic M, Martín-Alonso J, Barceló D (2008) Analysis and occurrence of pharmaceuticals, estrogens, progestogens and polar pesticides in sewage treatment plant effluents, river water and drinking water in the Llobregat river basin (Barcelona, Spain). J Hydrol 358:112–123. https://doi.org/10.1016/j.jhydrol.2008.05.030

Kvarnryd M, Grabic R, Brandt I, Berg C (2011) Early life progestin exposure causes arrested oocyte development, oviductal agenesis and sterility in adult *Xenopus tropicalis* frogs. Aquat Toxicol 103:18–24. https://doi.org/10.1016/j.aquatox.2011.02.003

Labadie P, Budzinski H (2005) Development of an analytical procedure for determination of selected estrogens and progestagens in water samples. Anal Bioanal Chem 381:1199–1205. https://doi.org/10.1007/s00216-005-3055-1

Liang YQ, Huang GY, Liu SS, Zhao JL, Yang YY, Chen XW, Tian F, Jiang YX, Ying GG (2015a) Long-term exposure to environmentally relevant concentrations of progesterone and norgestrel affects sex differentiation in zebrafish (*Danio rerio*). Aquat Toxicol 160:172–179. https://doi.org/10.1016/j.aquatox.2015.01.006

Liang YQ, Huang GY, Ying GG, Liu SS, Jiang YX, Liu S (2015b) Progesterone and norgestrel alter transcriptional expression of genes along the hypothalamic-pituitary-thyroid axis in zebrafish embryos-larvae. Comp Biochem Physiol C Toxicol Pharmacol 167:101–107. https://doi.org/10.1016/j.cbpc.2014.09.007

Liang YQ, Huang GY, Lin Z, Li J, Yang JW, Zhong LY, Ying GG (2018) Reproductive effects of synthetic progestin norgestrel in zebrafish (*Danio rerio*). Chemosphere 190:17–24. https://doi.org/10.1016/j.chemosphere.2017.09.127

Liu S, Ying GG, Zhao JL, Chen F, Yang B, Zhou LJ, Lai HJ (2011) Trace analysis of 28 steroids in surface water, wastewater and sludge samples by rapid resolution liquid chromatography-electrospray ionization tandem mass spectrometry. J Chromatogr A 1218:1367–1378. https://doi.org/10.1016/j.chroma.2011.01.014

Lopez de Alda MJ, Gil A, Paz E, Barcelo D (2002) Occurrence and analysis of estrogens and progestogens in river sediments by liquid chromatography–electrospray-mass spectrometry. Analyst 127:1299–1304. https://doi.org/10.1039/b207658f

Lorenz C, Contardo-Jara V, Pflugmacher S, Wiegand C, Nutzmann G, Lutz I, Kloas W (2011a) The synthetic gestagen levonorgestrel impairs metamorphosis in *Xenopus laevis* by disruption of the thyroid system. Toxicol Sci 123:94–102. https://doi.org/10.1093/toxsci/kfr159

Lorenz C, Contardo-Jara V, Trubiroha A, Krüger A, Viehmann V, Wiegand C, Pflugmacher S, Nützmann G, Lutz I, Kloas W (2011b) The synthetic gestagen levonorgestrel disrupts sexual development in *Xenopus laevis* by affecting gene expression of pituitary gonadotropins and gonadal steroidogenic enzymes. Toxicol Sci 124:311–319. https://doi.org/10.1093/toxsci/kfr241

Lorenz C, Opitza R, Trubiroha A, Lutz I, Zikova A, Kloas W (2016) The synthetic gestagen levonorgestrel directly affects gene expression in thyroid and pituitary glands of *Xenopus laevis* tadpoles. Aquat Toxicol 177:63–73. https://doi.org/10.1016/j.aquatox.2016.05.011

Lyman WJ, Reehl WF, Rosenblatt DH (1990) Handbook of chemical property estimation methods. American Chemical Society. pp. 7-4, 7-5, 8-12, Washington, DC

Maasz G, Zrinyi Z, Takacs P, Lovas S, Fodor I, Kiss T, Pirger Z (2017) Complex molecular changes induced by chronic progestogens exposure in roach, *Rutilus rutilus*. Ecotoxicol Environ Saf 139:9–17. https://doi.org/10.1016/j.ecoenv.2017.01.020

Matějíček D, Kubáň V (2007) High performance liquid chromatography/ion-trap mass spectrometry for separation and simultaneous determination of ethynylestradiol, gestodene, levonorgestrel, cyproterone acetate and desogestrel. Anal Chim Acta 588:304–315. https://doi.org/10.1016/j.aca.2007.02.028

Moermond CTA, Kase R, Korkaric M, Ågerstrand M (2016) CRED: criteria for reporting and evaluating ecotoxicity data. Environ Toxicol Chem 35:1297–1309. https://doi.org/10.1002/etc.3259

Narváez JF, Grant H, Gil VC, Porras J, Bueno Sanchez JC, Ocampo Duque LF, Sossa RR, Quintana-Castillo JC (2019) Assessment of endocrine disruptor effects of levonorgestrel and its photoproducts: Environmental implications of released fractions after their photocatalytic removal. J Hazard Mater 371:273–279. https://doi.org/10.1016/j.jhazmat.2019.02.095

OECD (Organization for Economic Cooperation and Development) (2004) Guidelines for testing of chemicals No. 202: *Daphnia magna* acute immobilization test. OECD, Paris

OECD (Organization for Economic Cooperation and Development) (2008) Guidelines for testing of chemicals No. 211: *Daphnia magna* reproduction test. OECD, Paris

O'Neil MJ, Heckelman PE, Koch CB, Roman KJ (2006) The Merck index: an encyclopedia of chemicals, drugs, and biologicals, CD-ROM encyclopedia, 14th edn. Merck/ Wiley, Hoboken

Oropesa AL, Soler F, Martins A, Guimarães L, Guilhermino L (2010) Acute and chronic toxicity of the progestin levonorgestrel to *Daphnia magna*. In: SETAC Europe 20th Annual Meeting. Sevilla (Spain)

Overturf MD, Overturf CL, Carty DR, Hala D, Huggett DB (2014) Levonorgestrel exposure to fathead minnows (*Pimephales promelas*) alters survival, growth, steroidogenic gene expression and hormone production. Aquat Toxicol 148:152–161. https://doi.org/10.1016/j.aquatox.2014.01.012

Perkins EJ, Ashauer R, Burgoon L, Conolly R, Landesmann B, Mackay C, Murphy CA, Pollesch N, Wheeler JR, Zupanic A, Scholz S (2019) Building and applying quantitative adverse outcome pathway models for chemical hazard and risk assessment. Environ Toxicol Chem 38:1850–1865

Petrovic M, Sole M, Lopez de Alda MJ, Barcelo D (2002) Endocrine disruptors in sewage treatment plants, receiving river waters, and sediments: integration of chemical analysis and biological effects on feral carp. Environ Toxicol Chem 21:2146–2156. https://doi.org/10.1002/etc.5620211018

Phillips A, Demarest DW, Hahn F, Wong F, McGuire JL (1990) Progestational and androgenic receptor binding affinities and *in vivo* activities of norgestimate and other progestins. Contraception 41:399–410. https://doi.org/10.1016/0010-7824(90)90039-X

Pietsch C, Neumann N, Knopf K, Wuertz S, Kloas W (2009) Progestogens cause immunosuppression of stimulated carp (*Cyprinus carpio* L.) leukocytes *in vitro*. Comp Biochem Physiol C Toxicol Pharmacol 150:16–24

Pietsch C, Neumann N, Preuer T, Kloas W (2011) *In vivo* treatment with progestogens causes immunosuppression of carp *Cyprinus carpio* leucocytes by affecting nitric oxide production and arginase activity. J Fish Biol 79:53–69. https://doi.org/10.1111/j.1095-8649.2011.02981.x

Pu C, Wu YF, Yang H, Deng AP (2008) Trace analysis of contraceptive drug levonorgestrel in wastewater samples by a newly developed indirect competitive enzyme-linked immunosorbent assay (ELISA) coupled with solid phase extraction. Anal Chim Acta 628:73–79. https://doi.org/10.1016/j.aca.2008.08.034

Qiao Y, Yang H, Wang B, Song J, Deng A (2009) Preparation and characterization of an immunoaffinity chromatography column for the selective extraction of trace contraceptive drug levonorgestrel from water samples. Talanta 80:98–103. https://doi.org/10.1016/j.talanta.2009.06.029

Runnalls TJ, Beresford N, Losty E, Scott AP, Sumpter JP (2013) Several synthetic progestins with different potencies adversely affect reproduction of fish. Environ Sci Technol 47:2077–2084. https://doi.org/10.1021/es3048834

Runnalls TJ, Beresford N, Kugathas S, Margiotta-Casaluci L, Scholze M, Scott AP, Sumpter JP (2015) From single chemicals to mixtures – Reproductive effects of levonorgestrel and ethinylestradiol on the fathead minnow. Aquat Toxicol 169:152–167. https://doi.org/10.1016/j.aquatox.2015.10.009

Säfholm M, Norder A, Fick J, Berg C (2012) Disrupted oogenesis in the frog *Xenopus tropicalis* after exposure to environmental progestin concentrations. Biol Reprod 86(4):1–7. https://doi.org/10.1095/biolreprod.111.097378

Säfholm M, Jansson E, Fick J, Berg C (2015) Mixture effects of levonorgestrel and ethinylestradiol: estrogenic biomarkers and hormone receptor mRNA expression during sexual programming. Aquat Toxicol 161:146–153. https://doi.org/10.1016/j.cbpc.2015.12.001

Säfholm M, Jansson E, Fick J, Berg C (2016) Molecular and histological endpoints for developmental reproductive toxicity in *Xenopus tropicalis*: levonorgestrel perturbs anti-Müllerian hormone and progesterone receptor expression. Comp Biochem Physiol C Toxicol Pharmacol 181-182:9–18. https://doi.org/10.1016/j.cbpc.2015.12.001

Šauer P, Stará A, Golovko O, Valentová O, Bořík A, Grabic R, Kocour Kroupová H (2018a) Two synthetic progestins and natural progesterone are responsible for most of the progestagenic activities in municipal wastewater treatment plant effluents in the Czech and Slovak republics. Water Res 137:64–71. https://doi.org/10.1016/j.watres.2018.02.065

Šauer P, Bořík A, Golovko O, Grabic R, Vojs Staňová A, Valentová O, Stará A, Šandová M, Kocour Kroupová H (2018b) Do progestins contribute to (anti-) androgenic activities in the aquatic environment? Environ Pollut 242:417–425. https://doi.org/10.1016/j.envpol.2018.06.104

Schmitt C, Streck G, Lamoree M, Leonards P, Brack W, de Deckere E (2011) Effect directed analysis of riverine sediments--the usefulness of *Potamopyrgus antipodarum* for *in vivo* effect confirmation of endocrine disruption. Aquat Toxicol 101(1):237–243. https://doi.org/10.1016/j.aquatox.2010.10.002

Scott PD, Bartkow M, Blockwell SJ, Coleman HM, Khan SJ, Lim R, McDonald JA, Nice H, Nugegoda D, Pettigrove V, Tremblay LA, Warne MSJ, Leusch FDL (2014) An assessment of endocrine activity in Australian rivers using chemical and in vitro analyses. Environ Sci Pollut Res 21:12951–12967. https://doi.org/10.1007/s11356-014-3235-7

Shen X, Chang H, Sun D, Wang L, Wu F (2018) Trace analysis of 61 natural and synthetic progestins in river water and sewage effluents by ultra-high performance liquid chromatography-tandem mass spectrometry. Water Res 133:142–152. https://doi.org/10.1016/j.watres.2018.01.030

Shi WJ, Zhao JL, Jiang YX, Huang GY, Liu YS, Zhang JN, Ying GG (2017) Transcriptional and histological alterations in gonad of adult zebrafish after exposure to the synthetic progestin norgestrel. Environ Toxicol Chem 36(12):3267–3276. https://doi.org/10.1002/etc.3894

Sun L, Yong W, Chu X, Lin JM (2009) Simultaneous determination of 15 steroidal oral contraceptives in water using solid-phase disk extraction followed by high performance liquid chromatography-tandem mass spectrometry. J Chromatogr A 1216:5416–5423. https://doi.org/10.1016/j.chroma.2009.05.041

Svensson J, Fick J, Brandt I, Brunström B (2013) The synthetic progestin levonorgestrel is a potent androgen in the three-spined stickleback (*Gasterosteus aculeatus*). Environ Sci Technol 47:2043–2051. https://doi.org/10.1021/es304305k

Svensson J, Fick J, Brandt I, Brunström B (2014) Environmental concentrations of an androgenic progestin disruptsthe seasonal breeding cycle in male three-spined stickleback (*Gasterosteus aculeatus*). Aquat Toxicol 147:84–91. https://doi.org/10.1016/j.aquatox.2013.12.013

Svensson J, Mustafa A, Fick J, Schmitz M, Brunström B (2016) Developmental exposure to progestins causes male bias and precocious puberty in zebrafish (*Danio rerio*). Aquat Toxicol 177:316–323. https://doi.org/10.1016/j.aquatox.2016.06.010

Thrupp TJ, Runnalls TJ, Scholze M, Kugathas S, Kortenkamp A, Sumpter JP (2018) The consequences of exposure to mixtures of chemicals: something from 'nothing' and 'a lot from a little' when fish are exposed to steroid hormones. Sci Total Environ 619-620:1482–1492. https://doi.org/10.1016/j.scitotenv.2017.11.081

Viglino L, Aboulfadl K, Prevost M, Sauve S (2008) Analysis of natural and synthetic estrogenic endocrine disruptors in environmental waters using online preconcentration coupled with LC-APPI-MS/MS. Talanta 76:1088–1096. https://doi.org/10.1016/j.talanta.2008.05.008

Viglino L, Prevost M, Sauve S (2011) High throughput analysis of solid-bound endocrine disruptors by LDTD-APCI-MS/MS. J Environ Monit 13:583–590. https://doi.org/10.1039/c0em00550a

Vulliet E, Cren-Olivé C (2011) Screening of pharmaceuticals and hormones at the regional scale, in surface and groundwaters intended to human consumption. Environ Pollut 159:2929–2934. https://doi.org/10.1016/j.envpol.2011.04.033

Vulliet E, Baugros JB, Flament-Waton MM, Grenier-Loustalot MF (2007) Analytical methods for the determination of selected steroid sex hormones and corticosteriods in wastewater. Anal Bioanal Chem 387:2143–2151. https://doi.org/10.1007/s00216-006-1084-z

Vulliet E, Wiest L, Baudot R, Grenier-Loustalot MF (2008) Multi-residue analysis of steroids at sub-ng/L levels in surface and ground-waters using liquid chromatography coupled to tandem mass spectrometry. J Chromatogr A 1210:84–91. https://doi.org/10.1016/j.chroma.2008.09.034

Vulliet E, Cren-Olivé C, Grenier-Loustalot MF (2011) Occurrence of pharmaceuticals and hormones in drinking water treated from surface waters. Environ Chem Lett 9:103–114. https://doi.org/10.1007/s10311-009-0253-7

Weizel A, Schlusener MP, Dierkes G, Ternes TA (2018) Occurrence of glucocorticoids, mineralocorticoids, and progestogens in various treated wastewater, rivers, and streams. Environ Sci Technol 52:5296–5307. https://doi.org/10.1021/acs.est.7b06147

Yarahmadi H, Duy SV, Hachad M, Dorner S, Sauvé S, Prévost M (2018) Seasonal variations of steroid hormones released by wastewater treatment plants to river water and sediments: Distribution between particulate and dissolved phases. Sci Total Environ 635:144–155. https://doi.org/10.1016/j.scitotenv.2018.03.370

Yu Q, Geng J, Zong X, Zhang Y, Xu K, Hu H, Deng Y, Zhao F, Ren H (2019) Occurrence and removal of progestagens in municipal wastewater treatment plants from different regions in China. Sci Total Environ 668:1191–1199. https://doi.org/10.1016/j.scitotenv.2019.02.327

Zeilinger J, Steger-Hartmann T, Maser E, Goller S, Vonk R, Länge R (2009) Effects of synthetic gestagens on fish reproduction. Environ Toxicol Chem 28(12):2663–2670. https://doi.org/10.1897/08-485.1

Zhao Y, Zhang K, Fent K (2018) Regulation of zebrafish (*Danio rerio*) locomotor behavior and circadian rhythm network by environmental steroid hormones. Environ Pollut 232:422–429. https://doi.org/10.1016/j.envpol.2017.09.057

Zucchi S, Castiglioni S, Fent K (2012) Progestins and antiprogestins affect gene expression in early development in zebrafish (*Danio rerio*) at environmental concentrations. Environ Sci Technol 46:5183–5185. https://doi.org/10.1021/es300231y

A Review on the Environmental Exposure to Airborne Manganese, Biomonitoring, and Neurological/Neuropsychological Outcomes

Ignacio Fernández-Olmo, Paula Mantecón, Bohdana Markiv, Laura Ruiz-Azcona, and Miguel Santibáñez

Contents

Abstract The occupational exposure to airborne manganese (Mn) has been linked for decades with neurological effects. With respect to its environmental exposure, the first reviews on this matter stated that the risk posed to human health by this kind of exposure was still unknown. Later, many studies have been developed to analyze the association between environmental Mn exposure and health effects, most of them including the measure of Mn in selected human biomarkers. This review aims at collecting and organizing the literature dealing with the environmental airborne Mn exposure (other routes of exposure were intentionally removed from this review), the biomonitoring of this metal in different body matrices (e.g., blood, urine, nails, hair), and the association between exposure and several adverse health

I. Fernández-Olmo (✉) · P. Mantecón · B. Markiv
Dpto. de Ingenierías Química y Biomolecular, Universidad de Cantabria, Santander, Cantabria, Spain
e-mail: fernandi@unican.es; paula.manteconc@alumnos.unican.es; bohdana.markiv@unican.es

L. Ruiz-Azcona · M. Santibáñez
Global Health Research Group, Dpto. Enfermería, Universidad de Cantabria-IDIVAL, Santander, Cantabria, Spain
e-mail: laura.ruiz@unican.es; miguel.santibanez@unican.es

P. de Voogt (ed.), *Reviews of Environmental Contamination and Toxicology Volume 254*, Reviews of Environmental Contamination and Toxicology 254,
https://doi.org/10.1007/398_2020_46

effects, such as, e.g., neurocognitive, neurodevelopmental, or neurobehavioral outcomes. From the different exposure routes, inhalation was the only one considered in this review, to take into account the areas influenced by industrial activities closely related to the Mn industry (ferromanganese and silicomanganese plants, Mn ore mines, and their processing plants) and by traffic in countries where a fuel additive, methylcyclopentadienyl manganese tricarbonyl (MMT), has been used for years. In these areas, high air Mn levels have been reported in comparison with the annual Reference Concentration (RfC) given by the US EPA for Mn, 50 ng/m^3. This review was performed using Scopus and MEDLINE databases with a keyword search strategy that took into account that each valid reference should include at least participants that were exposed to environmental airborne Mn and that were subjected to analysis of Mn in biomarkers or subjected to neurological/neuropsychological tests or both. Overall, 47 references matching these criteria were included in the discussion. Most of them report the measure of Mn in selected biomarkers ($N = 43$) and the assessment of different neurological outcomes ($N = 31$). A negative association is usually obtained between Mn levels in hair and some neurological outcomes, such as cognitive, motor, olfactory, and emotional functions, but not always significant. However, other biomarkers, such as blood and urine, do not seem to reflect the chronic environmental exposure to low/moderate levels of airborne Mn. Further studies combining the determination of the Mn exposure through environmental airborne sources and biomarkers of exposure and the evaluation of at least cognitive and motor functions are needed to better understand the effects of chronic non-occupational exposure to airborne Mn.

Keywords Airborne manganese · Biomarker · Environmental exposure · Neurological outcomes

1 Introduction

Manganese (Mn) is an oligoelement, vital to normal physiological function in humans (Chen et al. 2016). Manganese is needed in the daily diet of humans because it is required in the normal metabolism of carbohydrate, amino acid, lipid, and proteins (Erikson et al. 2007). As all the essential micronutrients, the deficiency and excess of Mn intake into the human body may result in adverse health effects (Mattison et al. 2017). Excessive exposure to Mn may occur in some occupational environments such as those encountered by welders and workers of the Mn industry, such as mines and smelters (Bowler et al. 2006a, 2006b; Erikson et al. 2007; Lucchini et al. 1999; Mergler et al. 1994; Roels et al. 1987). Manganism, a neurologic psychiatric disorder, comprising a variety of behavioral and motor impairments, such as difficulty with speech and walking (Chen et al. 2018), has been diagnosed in welders and in ferroalloy workers highly exposed to Mn (Bowler et al. 2006b; Lucchini et al. 1999). Although manganism resembles Parkinson's disease (PD), there is considerable evidence that Mn preferentially damages different areas of the brain from those that are affected in PD (Santamaria 2008).

A growing concern about the Mn environmental exposure of the general population has been raised in the last two decades. Several epidemiological studies have been carried out to assess the association between Mn exposure and adverse health effects in children and adults in areas exposed to Mn due to elevated Mn levels in air (Bowler et al. 2012; Haynes et al. 2015; Lucchini et al. 2012; Menezes-Filho et al. 2009a; Solís-Vivanco et al. 2009) and water (Bouchard et al. 2011; He et al. 1994; Khan et al. 2012). Most of these health outcomes will be discussed in this review; as a summary, excessive environmental Mn exposure can lead to adverse effects on the cognitive function (Bouchard et al. 2011; Bowler et al. 2015; Carvalho et al. 2014; Haynes et al. 2015, 2018; Kim et al. 2009; Lucchini et al. 2014; Menezes-Filho et al. 2011; Riojas-Rodríguez et al. 2010; Viana et al. 2014), mood changes (Mattison et al. 2017; Solís-Vivanco et al. 2009), and emotional function (Carvalho et al. 2018; Menezes-Filho et al. 2014; Rodrigues et al. 2018a) and other signs of toxicity on the motor function (Bowler et al. 2016; Kim et al. 2011; Mergler et al. 1999) or olfactory function (Guarneros et al. 2013; Lucchini et al. 2012, 2014). Despite the high number of studies dealing with the negative health effects of environmental Mn exposure, Chen et al. (2016) recommended that additional research into the associations between Mn exposure and human health is still needed.

Since Mn is naturally distributed in the earth crust, it is present in air, water, and soil mainly due to erosion (World Health Organization 2005). However, several anthropogenic sources release important amounts of Mn to the air, water, and soil; in addition, air Mn is deposited by wet and dry mechanisms, mainly near sources. The most important anthropogenic sources are Mn ore mining and processing, Mn alloy production plants (ferromanganese and silicomanganese), dry-cell battery plants, steel factories, iron foundries, fossil fuel combustion plants, and in some countries road traffic exhaust emissions due to the use of methylcyclopentadienyl Mn tricarbonyl (MMT) as an antiknock agent in unleaded fuel (US EPA 1984).

The main routes of environmental Mn exposure are inhalation of airborne Mn-bearing particles; ingestion of food, supplements, and water from wells with high Mn levels; accidental ingestion of contaminated soil and dust; and dermal contact. Far from airborne Mn emission sources, the annual air Mn levels are relatively low (10–30 ng/m^3) (World Health Organization 2005), and diet is the most important route of Mn intake. However, population living in the vicinity of Mn industrial sources is exposed to higher levels of air Mn; in these communities, Mn intake by inhalation of particulate matter is usually the major exposure route. Most of the epidemiological studies carried out in the general population to assess the adverse neurological effects have been performed in areas close to air Mn sources, such as Mn ore mines (Guarneros et al. 2013; Hernández-Bonilla et al. 2016; Riojas-Rodríguez et al. 2010; Solís-Vivanco et al. 2009; Torres-Agustín et al. 2013), Mn ore processing plants (Bowler et al. 2015, 2016; Haynes et al. 2018; Kornblith et al. 2018), Mn ferroalloys plants (Beuter et al. 1999; Bowler et al. 2012, 2015, 2016; Carvalho et al. 2014, 2018; Chiu et al. 2017; Haynes et al. 2015; Kim et al. 2011; Kornblith et al. 2018; Lucchini et al. 2012, 2014; Menezes-Filho et al. 2011, 2014; Rugless et al. 2014; Standridge et al. 2008; Viana et al. 2014), and urban areas affected by MMT emissions (Bolté et al. 2010; Lin et al. 2011). In such studies, the population living in a radius of 1–20 km is exposed to medium to high air Mn concentrations; however, in other studies where the major route of Mn intake is water, only small communities that use high Mn well water are highly exposed (Bouchard et al. 2011). In addition, the toxicity of Mn is lower when ingested than inhaled because excess ingested Mn is regulated in the gastrointestinal tract (Schwartz et al. 1986). For these reasons, this review was only focused on the environmental exposure to airborne Mn.

In the last decades, some reviews have dealt with the Mn exposure and its association with adverse health effects in the general population (Bjørklund et al. 2017; Chen et al. 2016; Coetzee et al. 2016; Hudnell 1999; Leonhard et al. 2019; Menezes-Filho et al. 2009b; Rodríguez-Barranco et al. 2013; Roels et al. 2012; Santamaria 2008; Zoni et al. 2007), some of them being focused on children, which is considered the most susceptible group to Mn exposure (Bjørklund et al. 2017; Coetzee et al. 2016; Leonhard et al. 2019; Menezes-Filho et al. 2009b; Rodríguez-Barranco et al. 2013). Most of these papers concluded that Mn exposure may cause neurobehavioral and neurological signs and symptoms.

Bjørklund et al. (2017) discussed the cognitive, behavioral, and motor function effects on children exposed to Mn; they concluded that highly exposed children performed more poorly in school and on neurobehavioral tests. Chen et al. (2016) reported that at-risk population showed poorer intellectual function, learning and memory, and mental development. Hudnell (1999) also suggested that the risk of Parkinson-like syndrome diagnosis may increase with continued Mn exposure and aging. Menezes-Filho et al. (2009b) from an analysis of 12 research articles concluded that higher exposure to Mn in children was associated with poorer cognitive functions and hyperactive behavior. Rodríguez-Barranco et al. (2013) found that 14 of 17 articles, obtained from a literature search based on assessing the health effects of the exposure to Mn, reported a significant negative effect on neurodevelopment or behavioral disorders in children. A meta-analysis performed from the data of these articles revealed that an increase on Mn exposure (measured as a 50% increase of Mn levels in hair) would be associated with a decrease of 0.7 points in the intelligence quotient (IQ) of children aged 6–13 years. Roels et al. (2012) also calculated an overall decrease in full IQ of 2.62 points for a tenfold increase of hair Mn (MnH) using the data of 617 children from 3 areas exposed to Mn (Mexico, Brazil, and Québec). Coetzee et al. (2016) also concluded that MnH is the most consistent biomarker of Mn exposure for children, finding an inverse association between MnH levels and IQ scores. Taking into account the different neurobehavioral and neuropsychological effects associated with Mn exposure identified from a literature review, Zoni et al. (2007) proposed a battery of tests and questionnaires to be applied in epidemiological studies including tests of motor functions, response speed, cognitive functions, intellectual abilities, mood, and symptoms. In a recent systematic literature review, Leonhard et al. (2019) concluded that the current evidence does not demonstrate a causal adverse effect of environmental exposure to Mn and child intelligence, although an inverse association between Mn exposure and children neurodevelopmental outcomes is suggested in some cross-sectional studies.

2 Methods: Literature Search Strategy

A scientific literature search was performed in March 2019 using Scopus and MEDLINE electronic databases. A keyword search strategy was performed to take into account that each valid reference should include at least participants that were exposed to environmental airborne Mn and that were subjected to analysis of Mn in biomarkers or subjected to neurological tests or both. Exclusion criteria included occupational studies, works with animals, and other routes of Mn exposure, such as water. Figure 1 shows the combination of keywords used in this literature search. Different neurological results were searched using the following keywords: neurobehavioral OR cognitive OR psychomotor OR "intelligence test" OR motor

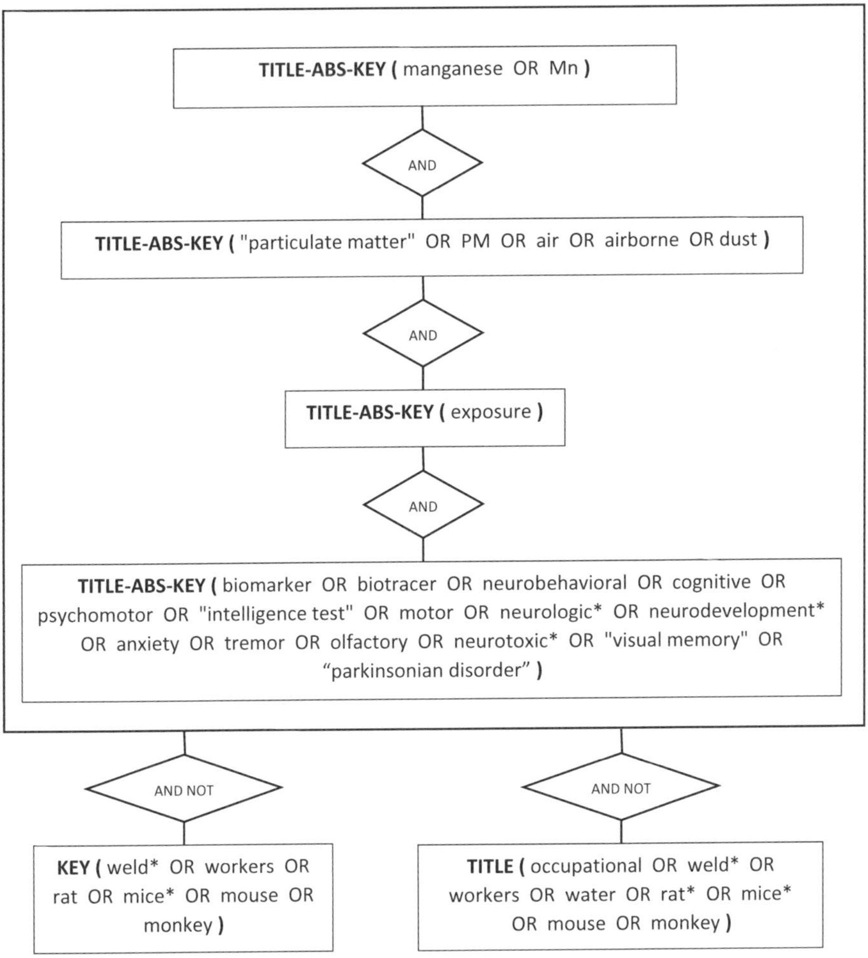

Fig. 1 Flow diagram of the search strategy

OR neurologic∗ OR neurodevelopment∗ OR anxiety OR tremor OR olfactory OR neurotoxic∗ OR "visual memory" OR "parkinsonian disorder". The exposure to environmental airborne Mn was in some references evaluated by measuring and/or modeling the particulate matter-bound Mn levels. Other references where air Mn concentrations were not explicitly evaluated were also considered if enough information about airborne Mn sources in the studied area was given. No restrictions about the age range of the participants were included; therefore, infants, children, and adults were considered in the search.

The results were limited to articles published in English. The references resulting from the search were first screened by title and abstract for relevance, and full texts of the selected references were then carefully reviewed. Additional references were also detected checking the reference list of the selected articles. Literature reviews were excluded from the analysis shown in the results and discussion sections; however, some reviews identified in the whole process were briefly discussed in the introduction section.

3 Results

Overall, 169 references matched all the search criteria in Scopus and 171 in MEDLINE databases and were assessed for eligibility. The analysis of titles and abstracts led to the selection of 35 articles in Scopus and 30 in MEDLINE (all of these were included in the Scopus search). Additional records (12) were identified through the references cited in the 35 articles selected from the primary search. Therefore, 47 full-text articles matched the selected criteria.

The information contained in the 47 selected references was organized in 4 tables. Table 1 summarizes the main contents of the selected articles showing the reference, country, type of study area, study population, and three qualitative columns to indicate: (1) if air Mn is evaluated; (2) if Mn is measured in biomarkers; and (3) if neurological outcomes are evaluated in the studied population. According to Table 1, 24 references report air Mn evaluation data by measuring or modeling the Mn levels in different environmental matrices (total suspended particles (TSP), PM10, PM2.5, dust, atmospheric deposition). These results are summarized in Table 2 and will be widely discussed later. Table 3 shows a summary of the results from 43 references that report Mn levels in different biomarkers such as blood, urine, hair, saliva, etc. The range of Mn levels in these biomarkers is discussed later. Finally, 31 references summarized in Table 4 report neurological outcomes that were assessed in the studied population. The cognitive and motor functions are evaluated in most of these articles by different tests that will be discussed later. Only 10 of 47 references provide results at the same time of the 3 main topics described in Tables 2, 3, and 4.

Table 1 Summary of study types

Reference	Country	Study area	Study population	Air Mn evaluation	Biomarkers of Mn exposure	Neurological outcomes
Baldwin et al. (1999)	Canada	Industrial: near a former Mn alloy plant	Adults	X	X	
Batterman et al. (2011)	South Africa	Urban/industrial	Children	X	X	
Beuter et al. (1999)	Canada	Industrial: near a former Mn alloy plant	Adults		X	X
Bolté et al. (2010)	Canada	Urban/rural	Adults	X	X	
Bowler et al. (2012)	United States	Industrial: near a Mn alloy plant	Adults	X	X	X
Bowler et al. (2015)	United States	Industrial: near a Mn alloy plant/near a Mn ore processing plant	Adults	X		X
Bowler et al. (2016)	United States	Industrial: near a Mn alloy plant/near a Mn ore processing plant	Adults	X		X
Carvalho et al. (2014)	Brazil	Industrial: near a Mn alloy plant	Children		X	X
Carvalho et al. (2018)	Brazil	Industrial: near a Mn alloy plant	Children		X	X
Chiu et al. (2017)	Italy	Industrial: near a former and an active Mn alloy plant	Children		X	X
Chung et al. (2015)	South Korea	Urban/industrial	Mothers/children		X	X
De Water et al. (2018)	Mexico	Urban	Mothers/children		X	X
Eastman et al. (2013)	Italy	Industrial: near a former and an active Mn alloy plant	Children		X	
Guarneros et al. (2013)	Mexico	Mining area, near a Mn ore processing plant	Adults		X	X
Haynes et al. (2010)	United States	Industrial: near a Mn alloy plant	Adults	X	X	
Haynes et al. (2012)	United States	Industrial: near a Mn alloy plant	Children	X	X	

(continued)

Table 1 (continued)

Reference	Country	Study area	Study population	Air Mn evaluation	Biomarkers of Mn exposure	Neurological outcomes
Haynes et al. (2015)	United States	Industrial: near a Mn alloy plant	Children		X	X
Haynes et al. (2018)	United States	Industrial: near a Mn ore processing plant	Children	X	X	X
Hernández-Bonilla et al. (2016)	Mexico	Mining area, near a Mn ore processing plant	Children	X	X	X
Kim et al. (2009)	South Korea	Urban	Children		X	X
Kim et al. (2011)	United States	Industrial: near a Mn alloy plant	Adults	X	X	X
Kornblith et al. (2018)	United States	Industrial: near a Mn alloy plant/near a Mn ore processing plant	Adults	X		X
Lin et al. (2011)	Taiwan	Road traffic	Newborns		X	
Lucas et al. (2015)	Italy	Industrial: near a former and an active Mn alloy plant	Children	X	X	
Lucchini et al. (2007)	Italy	Industrial: near a former and an active Mn alloy plant	Adults	X		X
Lucchini et al. (2012)	Italy	Industrial: near a former Mn alloy plant	Children	X	X	X
Lucchini et al. (2014)	Italy	Industrial: near a former Mn alloy plant	Adults	X	X	X
Menezes-Filho et al. (2009a, b)	Brazil	Industrial: near a Mn alloy plant	Children	X	X	
Menezes-Filho et al. (2011)	Brazil	Industrial: near a Mn alloy plant	Children		X	X
Menezes-Filho et al. (2014)	Brazil	Industrial: near a Mn alloy plant	Children		X	X
Mergler et al. (1999)	Canada	Industrial: near a former Mn alloy plant	Adults		X	X
Molina-Villalba et al. (2015)	Spain	Mining area/industrial	Children		X	

Montes et al. (2008)	Mexico	Mining area, near a Mn ore processing plant	Adults		X	
Ranzi et al. (2013)	Italy	Industrial/rural (near a municipal solid waste incinerator)	Adults		X	
Reis et al. (2015)	Portugal	Industrial	Adults	X	X	
Riojas-Rodríguez et al. (2010)	Mexico	Mining area, near a Mn ore processing plant	Children	X	X	X
Rodrigues et al. (2018a)	Brazil	Industrial: near a Mn alloy plant	Children		X	X
Rodrigues et al. (2018b)	Brazil	Industrial: near a Mn alloy plant	Children	X	X	
Rodríguez-Agudelo et al. (2006)	Mexico	Mining area, near a Mn ore processing plant	Adults	X	X	X
Rugless et al. (2014)	United States	Industrial: near a Mn alloy plant	Children		X	X
Santos-Burgoa et al. (2001)	Mexico	Mining area, near a Mn ore processing plant	Adults	X	X	
Solís-Vivanco et al. (2009)	Mexico	Mining area, near a Mn ore processing plant	Adults	X	X	X
Standridge et al. (2008)	United States	Industrial: near a Mn alloy plant	Adults		X	X
Torrente et al. (2005)	Spain	Urban/industrial	Children		X	X
Torres-Agustín et al. (2013)	Mexico	Mining area, near a Mn ore processing plant	Children	X	X	X
Viana et al. (2014)	Brazil	Industrial: near a Mn alloy plant	Adults		X	X
Vimercati et al. (2016)	Italy	Industrial	Adults		X	

Table 2 Summary of environmental airborne Mn evaluation

Reference	Region/city/town	Country	Study area	Evaluation of air Mn	Model	Sampler type	Matrix	Mean[a]	Minimum	Maximum	Units	Year	Sampling period
Baldwin et al. (1999)	Southwest Québec	Canada	Industrial: near a former Mn alloy plant	Measurement		Stationary	TSP	22	9	35	ng/m^3	1996	3 weeks
							PM10	12.5	7	19	ng/m^3		
Batterman et al. (2011)	Durban	South Africa	Urban	Measurement		Stationary	PM10	49	4.7	172.5	ng/m^3	2004/2005	9 months
							PM2.5	17	0.2	115.9	ng/m^3		
Bolté et al. (2010)	Montreal	Canada	Urban	Measurement		Stationary	Respirable, PM5	25	8	69	ng/m^3	–	3 days
	St-Philippe		Rural					5	2	15	ng/m^3		
Bowler et al. (2012)	Marietta	United States	Industrial: near a Mn alloy plant	Modeling	AERMOD		Air Mn	180	40	960	ng/m^3	1991/1995	5 years
Bowler et al. (2015, 2016)	Marietta	United States	Industrial: near a Mn alloy plant	Modeling	AERMOD		TSP	210	30	1,610	ng/m^3	2003/2013	11 years
							PM10	180	30	1,330	ng/m^3		
							PM2.5	50	7	340	ng/m^3		
	East Liverpool	United States	Industrial: near a Mn ore processing plant	Modeling	AERMOD		TSP	880	10	6,320	ng/m^3		
							PM10	310	5	2,210	ng/m^3		
							PM2.5	30	1	230	ng/m^3		
Haynes et al. (2010)	Marietta	United States	Industrial: near a Mn alloy plant	Modeling	AERMOD		Air Mn	130	10	18,130	ng/m^3	2006	1 year
Haynes et al. (2012)	Marietta	United States	Industrial: near a Mn alloy plant	Measurement		Personal (PEM)	PM2.5	8.1[b]	1.5	54.5	ng/m^3	2009/2010	7 months
						Stationary	PM2.5	11[b]	4.5	34.5	ng/m^3		
Haynes et al. (2018)	East Liverpool	United States	Industrial: near a Mn ore processing plant	Measurement (data from US EPA)		Stationary	TSP	433.3	110	1,010	ng/m^3	2003/2014	12 years
								433.3	90	1,040	ng/m^3	2013/2014	15 months

Hernández-Bonilla et al. (2016)	Tolago/ Chiconcoac	Mexico	Mining area, near a Mn ore processing plant	Measurement		Stationary	PM2.5	80	–	–	ng/m^3	2006	1 year
							PM10	470	–	–	ng/m^3		
	Agua blanca	Mexico	Control	Measurement		Stationary	PM2.5	30	–	–	ng/m^3		
							PM10	20	–	–	ng/m^3		
	Tolago/ Chiconcoac	Mexico	Mining area, near a Mn ore processing plant	Measurement		Stationary	PM2.5	10	–	–	ng/m^3	2013	1 year
							PM10	240	–	–	ng/m^3		
	Agua blanca	Mexico	Control	Measurement		Stationary	PM2.5	30	–	–	ng/m^3		
							PM10	20	–	–	ng/m^3		
Kim et al. (2011)	Marietta	United States	Industrial: near a Mn alloy plant	Modeling	AERMOD		Air Mn	180	40	960	ng/m^3	2001	1 year
Kornblith et al. (2018)	East Liverpool	United States	Industrial: near a Mn ore processing plant	Modeling	AERMOD		Air Mn	–	10	6,320	ng/m^3	–	10 years
	Marietta	United States	Industrial: near a Mn alloy plant					–	30	1,610	ng/m^3		
Lucas et al. (2015)	Bagnolo Mella	Italy	Industrial: near a Mn alloy plant	Measurement			Dust	11,700	487	183,000	μg/g	2010–2013	35 months
	Valcamonica	Italy	Industrial: near a former Mn alloy plant					1,350	407	8,240	μg/g		
	Garda Lake	Italy	Control					580	258	7,240	μg/g		
Lucchini et al. (2007)	Valcamonica	Italy	Industrial: near a Mn alloy plant	Measurement		Stationary	PM2.5	690[b]	200	1,800	ng/m^3	2001	–
Lucchini et al. (2012)	Valcamonica	Italy	Industrial: near a former Mn alloy plant	Measurement		Personal (PEM), n=125	PM10	49.5	1.24	516.7	ng/m^3	–	–
	Garda Lake	Italy	Control			Personal (PEM), n=64		27.4	5.3	85.9	ng/m^3	–	–

(continued)

Table 2 (continued)

Reference	Region/city/town	Country	Study area	Evaluation of air Mn	Model	Sampler type	Matrix	Mean[a]	Minimum	Maximum	Units	Year	Sampling period
Lucchini et al. (2014)	Valcamonica	Italy	Industrial: near a former Mn alloy plant	Measurement		Personal (PEM), n=153	PM10	26.41	1.99	103	ng/m^3	–	24 hours
						n=153	Soil	1,026	473	1,724	μg/g	–	–
	Garda Lake	Italy	Control	Measurement		Personal (PEM), n=101	PM10	20.96	2.44	68.1	ng/m3	–	24 hours
						n=102	Soil	421	313	549	μg/g	–	–
Menezes-Filho et al. (2009a, 2009b)	Simões Filho	Brazil	Industrial: near a Mn alloy plant	Measurement		Stationary	PM2.5	151	11	439	ng/m^3	2007	7 days
Reis et al. (2015)	Estarreja (Aveiro)	Portugal	Industrial	Measurement			Dust	265	186	472	μg/g	–	–
Riojas-Rodríguez et al. (2010)	Molango	Mexico	Mining area, near a Mn ore processing plant	Measurement		Stationary	PM10	130[c]	10	2,200	ng/m^3	2006–2007	–
	Agua blanca	Mexico	Control					20[c]	–	–	ng/m^3		
Rodrigues et al. (2018b)	Simões Filho	Brazil	Industrial: near a Mn alloy plant	Measurement			Bulk deposition	–	38.1	1,443.3	$\mu g/m^2$/day	-	37–41 days
Rodríguez-Agudelo et al. (2006)	Molango	Mexico	Mining area, near a Mn ore processing plant	Measurement		Personal (PEM), n=79	PM10	420	3	5,860	ng/m^3	2002	3 months
Santos-Burgoa et al. (2001)	Hidalgo State	Mexico	Mining area, near a Mn ore processing plant	Measurement		Stationary	PM10	70	30	100	ng/m^3	–	5 days
Solís-Vivanco et al. (2009)	Molango	Mexico	Mining area, near a Mn ore processing plant	Measurement		Personal (PEM), n=81	PM10	420	0	5,860	ng/m^3	–	–
Torres-Agustín et al. (2013)	Tolago/Chiconcoac	Mexico	Mining area, near a Mn ore processing plant	Measurement		Stationary	PM2.5	80[c]	20	240	ng/m^3	2007	14 days
	Agua blanca	Mexico	Control					20[c]	3	90	ng/m^3		7 days

[a]Arithmetic mean
[b]Geometric mean
[c]Median

Table 3 Summary of biomarkers of Mn exposure

Reference	Region/city/town	Country	Study area	Year	Population study	n	Biological matrix	Mean[a]	Minimum	Maximum	Units
Baldwin et al. (1999) and Beuter et al. (1999)	Southwest Québec	Canada	Industrial: near a former Mn alloy plant	1996	Adults	297	Blood	7.5	2.5	15.9	μg/L
Batterman et al. (2011)	Durban	South Africa	Urban	2002	Children	359	Blood	10.1	3	25	μg/L
Bolté et al. (2010)	Montreal	Canada	Urban	–	Adults (women)	5	Blood	8.4	5.4	10.8	μg/L
	St-Philippe	Canada	Rural	–				7.8	3.5	10.4	μg/L
Bowler et al. (2012)	Marietta	United States	Industrial: near a Mn alloy plant	2009	Adults	100	Blood	9.65	4.91	24.6	μg/L
	Mount Vernon	United States	Control			90		9.48	3.75	18.9	μg/L
Carvalho et al. (2014) and Menezes-Filho et al. (2014)	Simões Filho	Brazil	Industrial: near a Mn alloy plant	–	Children	70	Hair	14.6	0.52	55.74	μg/g
Carvalho et al. (2018)	Simões Filho	Brazil	Industrial: near a Mn alloy plant	2012	Children	70	Hair	14.6	0.52	55.74	μg/g
Chiu et al. (2017)	Valcamonica	Italy	Industrial: near a former Mn alloy plant	2013	Children	194	Prenatal dentine	0.43 10^{-4}	–	–	Mn/Ca ratio
							Blood	11	9.1	13	μg/L
Chung et al. (2015)	Seoul/Cheonan/Ulsan	South Korea	Urban/urban/industrial	2007/2009	Mothers	232	Blood	22.5	–	–	μg/L
De Water et al. (2018)	Mexico D.F.	Mexico	Urban	2007/2011	Mothers	15	Blood	17	2.7	41.1	μg/L

(continued)

Table 3 (continued)

Reference	Region/city/ town	Country	Study area	Year	Population study	n	Biological matrix	Mean[a]	Minimum	Maximum	Units
Eastman et al. (2013)	Bagnolo Mella	Italy	Industrial: near a Mn alloy plant	–	Children	44	Hair	0.134	0.011	0.736	μg/g
	Valcamonica	Italy	Industrial: near a former Mn alloy plant	–		41		0.07	0.025	0.642	μg/g
	Garda Lake	Italy	Control	–		36		0.06	0.023	0.344	μg/g
Guarneros et al. (2013)	Tolago/ Chiconcoac	Mexico	Mining area, near a Mn ore processing plant	2011	Adults	25	Hair	9.73	–	–	μg/g
	Calnali	Mexico	Control			23		1.01	–	–	μg/g
Haynes et al. (2010)	Marietta	United States	Industrial: near a Mn alloy plant	2006	Adults	135	Blood	9.12	1.8	22	μg/L
						73	Hair	5.8	0.64	41.1	μg/g
Haynes et al. (2012)	Marietta	United States	Industrial: near a Mn alloy plant	2009/ 2010	Children	38	Blood	9.5	5	14.4	μg/L
						38	Hair	0.47	0.085	1.25	μg/g
Haynes et al. (2015)	Marietta/ Cambridge	United States	Industrial: near a Mn alloy plant	2008/ 2013	Children	327	Blood	9.67	–	–	μg/L
						370	Hair	0.4165	–	–	μg/g
Haynes et al. (2018)	East Liverpool	United States	Mining area, near a Mn ore processing plant	2013/ 2014	Children	66	Blood	10.06	5.9	18.4	μg/L
						98	Hair	0.360	0.017	15.967	μg/g
Hernández-Bonilla et al. (2016)	Tolago/ Chiconcoac	Mexico	Mining area, near a Mn ore processing plant	2006/ 2013	Children	148	Hair	5.25[b]	–	–	μg/g
	Agua blanca	Mexico	Control			119		0.55[b]	–	–	μg/g
Kim et al. (2009)	Seoul, Seong-nam, Ulsan, Yeoncheon	South Korea	Urban	2007	Children	261	Blood	14.3	5.3	29.02	μg/L

Kim et al. (2011)	Marietta	United States	Industrial: near a Mn alloy plant	–	Adults	100	Blood	9.65	4.91	24.6	µg/L
	Mount Vernon	United States	Control	–		90		9.48	3.75	18.9	µg/L
Lin et al. (2011)	Taipei, Yune-Lin, Chia-Yi, Tainan, Kaohsiung, Taitung	Taiwan	Road traffic	2004/2005	Newborns	1,343	Cord blood	47[b]	45	49.9	µg/L
Lucas et al. (2015)	Brescia	Italy	Mean between reference and industrial area (near a Mn alloy plant)	–	Children	546	Blood	11.3	4	34.33	µg/L
						249	Saliva	21	0.16	305	µg/L
						501	Hair	0.143	0.011	1.13	µg/g
						207	Fingernail	0.525	0.006	8.28	µg/g
Lucchini et al. (2012)	Valcamonica	Italy	Industrial: near a former Mn alloy plant	–	Adolescents	151	Blood	10.99	4	21.6	µg/L
						125	Hair	0.16	0.02	1.27	µg/g
						151	Urine	0.22	0.1	7.6	µg/L
	Garda Lake	Italy	Control	–		148	Blood	11.24	6	24.1	µg/L
						133	Hair	0.18	0.02	3.45	µg/g
						150	Urine	0.16	0.1	1.5	µg/L
Lucchini et al. (2014)	Valcamonica	Italy	Industrial: near a former Mn alloy plant	–	Elderly	146	Blood	8.4	3.6	19.5	µg/L
						146	Urine	0.3	0.1	6	µg/L
	Garda Lake	Italy	Control	–	Elderly	92	Blood	10.2	3.6	21.6	µg/L
						93	Urine	0.4	0.1	9.4	µg/L
Menezes-Filho et al. (2009a, 2009b)	Simões Filho	Brazil	Industrial: near a Mn alloy plant	2007	Children	109	Hair	15.2	1.1	95.5	µg/g
	Capiarara	Brazil	Control			43		1.37	0.39	5.58	µg/g
Menezes-Filho et al. (2011)	Simões Filho	Brazil	Industrial: near a Mn alloy plant	2007	Children	70	Blood	8.2	2.7	23.4	µg/L
						83	Hair	5.83	0.1	86.68	µg/g
					Mothers	77	Hair	3.5	0.1	77.45	µg/g

(continued)

Table 3 (continued)

Reference	Region/city/town	Country	Study area	Year	Population study	n	Biological matrix	Mean[a]	Minimum	Maximum	Units
Menezes-Filho et al. (2014)	Simões Filho	Brazil	Industrial: near a Mn alloy plant	–	Mothers	34	Hair	10.71	0.62	44.61	μg/g
Mergler et al. (1999)	Southwest Québec	Canada	Industrial: near a former Mn alloy plant	1996	Adults	273	Blood	7.3	2.5	15.9	μg/L
Molina-Villalba et al. (2015)	Huelva	Spain	Mining area/industrial	2012	Children	261	Urine	0.12	–	6.127	μg/L
						220	Hair	0.263	–	2.624	μg/g
Montes et al. (2008)	Molango	Mexico	Mining area, near a Mn ore processing plant	–	Adults	230	Blood	9.68	5	26.5	μg/L
Ranzi et al. (2013)	Modena	Italy	Industrial/rural (near a municipal solid waste incinerator)	2010	Adults	65	Urine	0.69	–	–	μg/L
			Control			103		0.59			μg/L
Reis et al. (2015)	Estarreja (Aveiro)	Portugal	Industrial	–	Adults	21	Toenail	0.34	0.09	1	μg/g
					Children	6		0.7	0.14	2.25	μg/g
Riojas-Rodríguez et al. (2010)	Molango	Mexico	Mining area, near a Mn ore processing plant	2006/2007	Children	79	Blood	9.71[b]	5.5	18	μg/L
			Mining area, near a Mn ore processing plant			79	Hair	12.13[b]	4.2	48	μg/g
	Agua blanca	Mexico	Control	2006/2007	Children	93	Blood	8.22[b]	5	14	μg/L
			Control			93	Hair	0.57[b]	0.06	3.64	μg/g
Rodrigues et al. (2018a)	Simões Filho	Brazil	Industrial: near a Mn alloy plant	–	Children	225	Hair	0.73[c]	0.16	8.79	μg/g
							Toenail	0.84[c]	0.15	9.29	μg/g
							Blood	8.98[c]	1.51	40.43	μg/L

Rodrigues et al. (2018b)	Simões Filho	Brazil	Industrial: near a Mn alloy plant	–	Children	163	Hair	0.66	0.16	8.79	μg/g
						157	Toenail	0.86	0.15	13.3	μg/g
Rodríguez-Agudelo et al. (2006)	Molango	Mexico	Mining area, near a Mn ore processing plant	2002	Adults	288	Blood	10.16	5	31	μg/L
Rugless et al. (2014)	Marietta	United States	Industrial: near a Mn alloy plant	2009/2010	Children	48	Blood	10	6.1	18.8	μg/L
						48	Hair	0.77	0.1	7.4	μg/g
Santos-Burgoa et al. (2001)	Hidalgo State	Mexico	Mining area, near a Mn ore processing plant	–	Adults	73	Blood	17.71	7.5	88	μg/L
Solís-Vivanco et al. (2009)	Molango	Mexico	Mining area, near a Mn ore processing plant	–	Adults	288	Blood	10.16	5	31	μg/L
Standridge et al. (2008)	Marietta	United States	Industrial: near a Mn alloy plant	2006	Adults	28	Blood	9.4	4.2	21.7	μg/L
						29	Hair	4.4	1.2	12.4	μg/g
Torrente et al. (2005)	Tarragona	Spain	Industrial	–	Children	54	Hair	0.18	<LD	1.97	μg/g
			Urban			45		0.26	<LD	6	μg/g
Torres-Agustín et al. (2013)	Tolago/ Chiconcoac	Mexico	Mining area, near a Mn ore processing plant	2007	Children	79	Blood	10.04	5.5	18	μg/L
							Hair	14.2	4.2	48	μg/g
	Agua blanca	Mexico	Control	2007	Children	95	Blood	8[c]	5	14	μg/L
							Hair	0.73	0.06	3.6	μg/g
Viana et al. (2014)	Cotegipe-Simões Filho	Brazil	Industrial: near a Mn alloy plant	2010	Adults	42	Hair	2.7	0.6	44.6	μg/g
	Santa Luzia-Simões Filho	Brazil				47		10.5	0.9	42	μg/g
	Cotegipe-Simões Filho	Brazil				42	Axillary hair	5.8	3.8	17.2	μg/g
	Santa Luzia-Simões Filho	Brazil				47		21.8	4.4	85.6	μg/g

(continued)

Table 3 (continued)

Reference	Region/city/ town	Country	Study area	Year	Population study	n	Biological matrix	Mean[a]	Minimum	Maximum	Units
	Cotegipe-Simões Filho	Brazil				42	Fingernail	4	0.7	16.1	µg/g
	Santa Luzia-Simões Filho	Brazil				47		6.5	1.1	22.2	µg/g
	Cotegipe-Simões Filho	Brazil				42	Saliva	3	0.4	43.3	µg/L
	Santa Luzia-Simões Filho	Brazil				47		3.7	0.6	81.6	µg/L
Vimercati et al. (2016)	Taranto	Italy	Industrial	2010/ 2012	Adults	279	Urine	2.7	–	–	µg/L

[a]Arithmetic mean
[b]Geometric mean
[c]Median

Table 4 Summary of studies about neurological/neuropsychological outcomes in airborne Mn-exposed population

Reference	Region/city/town	Country	Study area	Year	Neurodevelopmental outcome	Specific outcomes
Beuter et al. (1999)	Southwest Québec	Canada	Industrial: near a former Mn alloy plant	–	Motor function	Eye-hand coordination, diadochokinesia
					Tremor	Acceleration of postural tremor
Bowler et al. (2012)	Marietta/Mount Vernon (control)	United States	Industrial: near a Mn alloy plant/control	2008/2009	Mood	Anxiety, mood
					Motor function	Motor symptoms, manual dexterity and motor speed, strength manipulative function, tactile manipulative function
					Cognitive function	Capacity for verbal concept formation, mental effort
Bowler et al. (2015)	Marietta/East Liverpool	United States	Industrial: near a Mn alloy plant/near a Mn ore processing plant	2009/2011	Cognitive function	Cognitive flexibility and executive functioning, information processing speed, memory (Neuropsychological Assessment Battery, NAB), working memory/attention and concentration/learning, visuospatial memory, visuomotor tracking speed, verbal skills, mental effort
Bowler et al. (2016)	Marietta/East Liverpool	United States	Industrial: near a Mn alloy plant/near a Mn ore processing plant	2009/2011	Motor function	Manual dexterity and motor speed, strength manipulative function, tactile manipulative function
					Tremor	Bilateral hand tremor
Carvalho et al. (2014)	Simões Filho	Brazil	Industrial: near a Mn alloy plant	–	Cognitive function	Intelligence, verbal memory, attention deficit, cognitive flexibility
Carvalho et al. (2018)	Simões Filho	Brazil	Industrial: near a Mn alloy plant	2012	Cognitive function	Inhibitory control and cognitive flexibility, verbal fluency, verbal memory, and learning
					Emotional function	Attention deficit and hyperactivity
					Motor function	Tactile manipulative function

(continued)

Table 4 (continued)

Reference	Region/city/town	Country	Study area	Year	Neurodevelopmental outcome	Specific outcomes
Chiu et al. (2017)	Valcamonica	Italy	Industrial: near a former Mn alloy plant	2013	Tremor	Tremor
Chung et al. (2015)	Seoul, Cheonan, Ulsan	South Korea	Urban/urban/industrial	2007/2011	Motor function	Motor abilities
					Cognitive function	Mental development
De Water et al. (2018)	Mexico D.F.	Mexico	Urban	2007/2011	Emotional function	Brain intrinsic functional connectivity (iFC)
Guarneros et al. (2013)	Tolago/Chiconcoac, Molango district// Calnali (control)	Mexico	Mining area, near a Mn ore processing plant/control	2011	Olfactory function	Odor threshold, discrimination, identification and overall performance
Haynes et al. (2015)	Marietta/Cambridge (control)	United States	Industrial: near a Mn alloy plant	2008/2013	Cognitive function	Full-scale IQ, perceptual reasoning, processing speed, working memory, verbal comprehension
Haynes et al. (2018)	East Liverpool	United States	Industrial: near a Mn ore processing plant	2013/2014	Cognitive function	Full-scale IQ, perceptual reasoning, processing speed, working memory, verbal comprehension
Hernández-Bonilla et al. (2016)	Molango/Agua Blanca	Mexico	Mining area, near a Mn ore processing plant/control	2006	Cognitive function	Visuoperception and short-term visual memory
					Motor function	Manual dexterity and motor speed
Kim et al. (2009)	Seoul, Seongnam, Ulsan, Yeoncheon	South Korea	Urban	2007	Cognitive function	Intelligence
Kim et al. (2011)	Marietta/Mount Vernon (control)	United States	Industrial: near a Mn alloy plant/control	2009	Motor function	Motor symptoms
					Tremor	Postural sway
Kornblith et al. (2018)	Marietta/East Liverpool	United States	Industrial: near a Mn alloy plant/near a Mn ore processing plant	2009/2011	Motor function	Motor symptoms
					Tremor	Tremor
					Cognitive function	Executive functioning: verbal capacity, response inhibition, visual tracking, visual planning, working memory

Lucchini et al. (2007)	Valcamonica	Italy	Industrial: near a former Mn alloy plant	2001	Motor function	Parkinsonian disorders
	Bagnolo Mella		Industrial: near a Mn alloy plant			
Lucchini et al. (2012)	Valcamonica/Garda Lake (control)	Italy	Industrial: near a former Mn alloy plant/control	–	Motor function	Motor coordination, manual dexterity and motor speed, psychomotor response, hand dexterity and perceptual speed
					Tremor	Tremor
					Olfactory function	Odor identification
Lucchini et al. (2014)	Valcamonica/Garda Lake (control)	Italy	Industrial: near a former Mn alloy plant/control	–	Cognitive function	Long-term verbal memory, clear-thinking ability, ability of spatial planning in a visualmotor track and the ability to quickly switch from a numeric stimulus to an alphabetical one
					Motor function	Motor speed
						Psychomotor speed
					Tremor	Tremor
					Olfactory function	Odor identification
Menezes-Filho et al. (2011)	Simões Filho: Cotegipe	Brazil	Industrial: near a Mn alloy plant	2008	Cognitive function	Intelligence
Menezes-Filho et al. (2014)	Simões Filho: Cotegipe/Santa Luzia	Brazil	Industrial: near a Mn alloy plant	–	Emotional function	Externalizing (disruptive and aggressive) behaviors, internalizing (withdrawn, somatic complaints, and anxious/depressed scales) behaviors
					Cognitive function	Intelligence in parents or caregivers

(continued)

Table 4 (continued)

Reference	Region/city/town	Country	Study area	Year	Neurodevelopmental outcome	Specific outcomes
Mergler et al. (1999)	Southwest Québec	Canada	Industrial: near a former Mn alloy plant	1996	Cognitive function	Attention and reaction time, cognitive flexibility, auditory learning, recall, delayed recall, visual recognition, delayed recall, visual memory, auditory recall and attention, cognitive flexibility, verbal naming
					Motor function	Proximal/distal pointing, forearm rotation, grapho-motor speed, manual dexterity and motor speed
					Sensorial function	Visual acuity, near visual contrast sensitivity, olfactory perception threshold
					Tremor	Arm and hand tremor, postural sway
Riojas-Rodríguez et al. (2010)	Molango/Agua Blanca	Mexico	Mining area, near a Mn ore processing plant/control	2006/2007	Cognitive function	Intelligence
Rodrigues et al. (2018a)	Simões Filho	Brazil	Industrial: near a Mn alloy plant	–	Cognitive function	Intelligence
					Emotional function	Externalizing (disruptive and aggressive) behaviors, internalizing (withdrawn, somatic complaints, and anxious/depressed scales) behaviors, attention problems
Rodríguez-Agudelo et al. (2006)	Molango	Mexico	Mining area, near a Mn ore processing plant	2002	Motor function	Speed and coordination with one hand, ability to imitate motor movement following a visual cue, simple and complex verbal control, intellectual-imaginative motor aptitude, oral movements, quality of drawing reproduction, conflictive and choice reactions

Rugless et al. (2014)	Marietta	United States	Industrial: near a Mn alloy plant	2009/ 2010	Motor function	Sway length, sway area
Solís-Vivanco et al. (2009)	Molango	Mexico	Mining area, near a Mn ore processing plant	–	Cognitive function	General cognitive state, attention problems, verbal memory, semantic fluency, phonological fluency, construction, visual memory, visual memory coding
					Mood	Beck depression, geriatric depression
Standridge et al. (2008)	Marietta	United States	Industrial: near a Mn alloy plant	2006	Motor function	Sway length, sway area
Torrente et al. (2005)	Tarragona-industrial zone/downtown (control)	Spain	Industrial: near a petrochemical plant	–	Cognitive function	Cognitive impairment, sustained attention, visuospatial capability, abstract reasoning
Torres-Agustín et al. (2013)	Molango/Agua Blanca	Mexico	Mining area, near a Mn ore processing plant/control	–	Cognitive function	Immediate recall, level of learning, immediate memory span, delayed recall, recognition accuracy
Viana et al. (2014)	Simões Filho: Cotegipe/Santa Luzia	Brazil	Industrial: near a Mn alloy plant	2010	Cognitive function	Intelligence, verbal memory immediate recall, selective and sustained visual attention, processing speed, cognitive flexibility, verbal working memory, visual working memory
					Motor function	Manual dexterity

Table 1 shows that these kinds of studies are mainly carried out in areas near industrial activities, such as closed Mn alloys plants (Canada, Italy), active Mn alloys plants (USA, Brazil, Italy), Mn ore processing plants (USA), and Mn ore mining activity (Mexico). Other works were also realized in urban areas, where MMT can be used as gasoline additive and in mixed urban/industrial areas. Most of these works are cross-sectional studies; only a few ones are cohort studies. The study population is composed of children (25 references, 2 of them in combination with their mothers) and adults (22 references).

3.1 Inhalation of Manganese: Evaluation of Airborne Manganese Levels

Table 2 summarizes the results of the evaluation of airborne Mn levels in the studied literature. Airborne Mn is usually evaluated by measuring or modeling the Mn concentration in different PM matrices, such as TSP (in some references "air Mn" is used), PM10, PM2.5, and respirable PM (PM4–5). Although stationary high- and low-volume samplers are usually used to collect PM samples, personal exposure samplers can also be used in cross-sectional studies (Haynes et al. 2012; Lucchini et al. 2012, 2014; Rodríguez-Agudelo et al. 2006; Solís-Vivanco et al. 2009). When air Mn is evaluated by modeling, relatively simple models such as AERMOD are used (Bowler et al. 2015, 2016; Haynes et al. 2010; Kim et al. 2011; Kornblith et al. 2018); this allows to estimate the air Mn levels at different geographical locations. More complex models such as SCIPUFF and CALPUFF have been also used in the literature to model air Mn levels, but they are not used in this kind of cross-sectional studies (Carter et al. 2015; Otero-Pregigueiro and Fernández-Olmo 2018).

The measured or modeled values are usually compared with some guidelines given by different organizations. Thus, the World Health Organization (WHO) recommends an annual average guideline of 150 ng Mn/m^3 (World Health Organization 2005). The US EPA (1993) established a Reference Concentration (RfC) of 50 ng/m^3 (annual basis) based on the lowest/no observable adverse effect level (LOAEL/NOAEL) analysis of an occupational study performed by Roels et al. (1992). A reference exposure level (REL) of 90 ng/m^3 was developed by California EPA's Office of Environmental Health Hazard Assessment (OEHHA) to protect the general public chronic exposure to airborne Mn (Winder et al. 2010). Later, the Agency for Toxic Substances and Disease Registry (ATSDR) developed a minimum risk level (MRL) for Mn of 300 ng/m^3 (ATSDR 2012); this value was also based on the same occupational study (Roels et al. 1992) but using updated dose-response modeling methodology and considering more recent pharmacokinetic findings. However, some authors consider that current regulatory air Mn guidelines are extremely conservative because of the nonlinear biological response to Mn exposure and that the new chronic guideline should be between 1 and 10 μg/m^3 (Bailey et al. 2009; Gentry et al. 2017). A summary of the current inhalation health effect Mn reference values for the general public can be found in US EPA (2012). This report stated that the particle size fraction where Mn should be evaluated is the respirable fraction (approximately PM5.0); however, only one of the studied references collected samples of the respirable fraction (Bolté et al. 2010). With respect to the European Union, no limit/threshold value has been established for Mn to date. According to Chen et al. (2016) and based on the potential neurological outcomes derived from the exposure to air Mn, there is a critical need to regulate Mn exposure.

The Mn levels in other environmental matrices closely related to the air Mn are also measured in some references as indicative of the Mn environmental exposure: atmospheric deposition (Rodrigues et al. 2018b), dust (Lucas et al. 2015; Reis et al. 2015), and soil (Lucchini et al. 2014).

To show visually the range of air Mn levels measured or modeled reported in the selected references, box plot diagrams of the average and maximum values given in these references were prepared (see Figs. 2 and 3); these values are compared with some of the guidelines discussed above. A broad distribution of the average Mn values can be observed in Fig. 2; the lowest values correspond to those measured in reference or control areas (i.e., non-exposed to air Mn sources). In areas located near industrial sources of Mn, the average values are in general higher than the guidelines values, reaching a maximum average value in TSP of 880 ng/m^3 in East Liverpool (USA) near a Mn ore handling and processing plant (Bowler et al. 2015); in PM10, Rodríguez-Agudelo et al. (2006) report an average value of 420 ng/m^3 near a mining district in Molango, Mexico; finally, the highest Mn average values in PM2.5 were found near ferromanganese plants in Simoes Filho, Brazil (151 ng/m^3, arithmetic mean), and in Valcamonica, Italy (690 ng/m^3, geometric mean) (Lucchini et al.

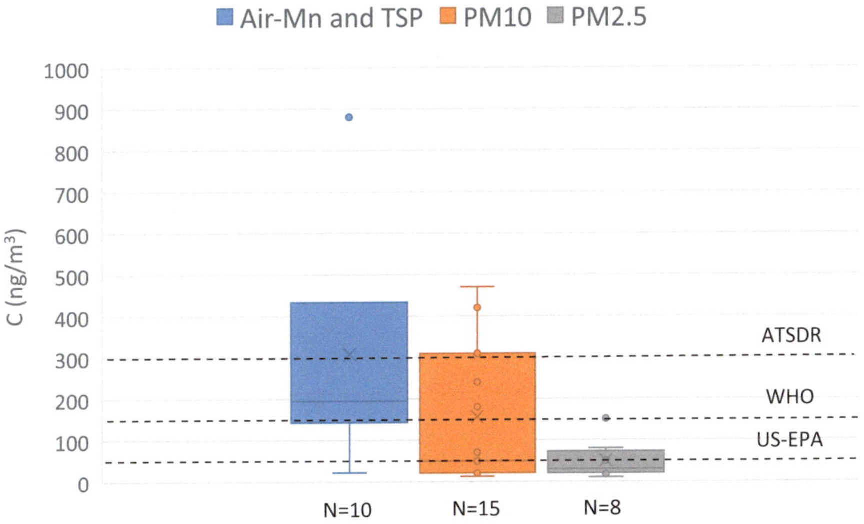

Fig. 2 Box plot of the average air Mn levels reported in the studied literature

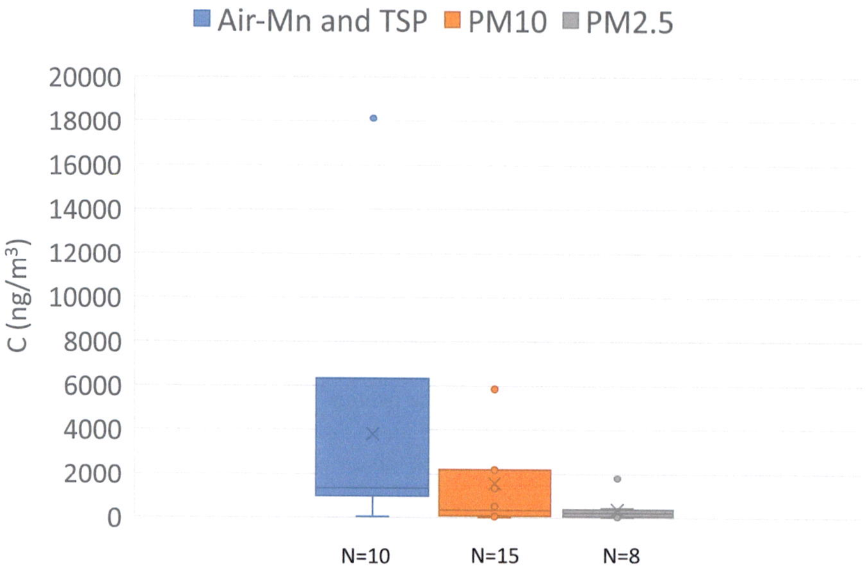

Fig. 3 Box plot of the maximum air Mn levels reported in the studied literature

2007; Menezes-Filho et al. 2009a). The maximum daily values can reach extremely high peaks as shown in Fig. 3: 18,130 ng/m^3 in TSP (Haynes et al. 2010), 5,860 ng/m^3 in PM10 (Rodríguez-Agudelo et al. 2006), and 1,800 ng/m^3 in PM2.5 (Lucchini et al. 2007).

3.2 *Biomarkers of Environmental Exposure to Manganese*

A biomarker of exposure can be defined as any measurable biological parameter that indicates levels of exposure to a given pollutant (Zheng et al. 2011). The selection of an appropriate biomarker of Mn exposure has been widely discussed in the literature. Lucchini et al. (2018) in the editorial published in the special issue derived from the International Manganese 2016 Conference recommended additional research on Mn biomarkers to understand the type of exposure and exposure windows that better reflected by each candidate biomarkers. According to Coetzee et al. (2016), there is still no consensus about the best biomarker of Mn exposure that better defines dose-effect relationships, probably because some mechanisms of Mn absorption and regulation are not completely known. This is particularly true when inhaling Mn; thus, although higher levels of Mn in blood of individuals exposed to high concentrations of Mn in workplaces are found, it is difficult to quantify how the pulmonary uptake of Mn-bearing particles contributes to the overall increase of blood Mn (MnB) (Roth 2006).

Even at occupational level, there are some discrepancies in the literature about the selection of appropriate biomarkers of Mn exposure. For example, Ellingsen et al. (2003) from a cross-sectional study developed with workers of a ferroalloy plant reported no association between inhalable Mn and MnB but a weak association with respirable Mn. The strongest association was found between urinary Mn (MnU) and the Mn-soluble fractions of respirable aerosol. However, other authors consider that MnU is not a good biomarker since Mn excreted in the urine only represents a small fraction of eliminated Mn (Bader et al. 1999; Zheng et al. 2011). Smith et al. (2007) found positive associations between MnB and total air Mn levels in ferromanganese plant workers exposed to low (median of 420 ng/m^3) and moderate (median of 4,200 ng/m^3) air Mn, but not in those workers exposed to very high air Mn levels (median of 292 μg/m^3); these authors stated that urine and plasma are not good biomarkers of Mn occupational exposure. Zheng et al. (2011) considered that MnB is a poor biomarker of Mn exposure because of the much shorter half-time of Mn in blood with respect to tissue and intracellular compartments.

With respect to the environmental exposure to Mn, the much lower pollutant doses make it more difficult to select a biomarker of exposure that better reflects the long-term exposure to different environmental sources. Coetzee et al. (2016) after reviewing the use of different biomarkers in children concluded that MnH was the most consistent and valid biomarker of Mn exposure. Although the mechanisms of Mn uptake into hair are not well-known (Bouchard et al. 2011), hair is a good candidate for long-term chronic exposure to Mn, since it grows slowly (Viana et al. 2014). The hair growth rate of approximately 1 cm per month provides an exposure estimate of 1–6 months (Haynes et al. 2015). The slow growth rate of nails also allows to consider it as a good biomarker (Viana et al. 2014). Ntihabose et al. (2018) also stated that hair and toenails are promising biomarkers of low-level chronic exposure to Mn in children.

Due to the relatively short life of Mn in blood, MnB does not represent the long-term accumulation in the body (Aschner et al. 2007; Bouchard et al. 2011), but it is likely to reflect recent exposures to Mn, even in environmental exposure to high levels of air Mn. For example, although the study reported by Smith et al. (2007), where positive associations between BMn and total air Mn levels were found, was carried out in an occupational environment, the concentration range of low-exposed workers was in the same order or magnitude of environmental Mn levels measured near manganese alloy plants and mining districts (i.e., 400 ng/m^3).

Table 3 summarizes the levels of Mn in biomarkers used in the selected literature; blood and hair are the main biological matrices used in these articles (27 and 23 references, respectively); other biomarkers are urine ($N = 5$), toenails ($N = 3$), fingernails ($N = 2$), saliva ($N = 2$), axillary hair ($N = 1$), umbilical cord blood ($N = 1$), and dentine ($N = 1$).

The average and maximum levels of MnB, MnH and MnU reported in the studied literature are plotted in Figs. 4 and 5, respectively. Figure 4 shows the narrow distribution of MnB in comparison with the wider distribution of MnU and mainly MnH (the box represents the range between the first and the third quartile). The median (50th percentile) of MnB is 9.68 μg/L, which agrees with the median found

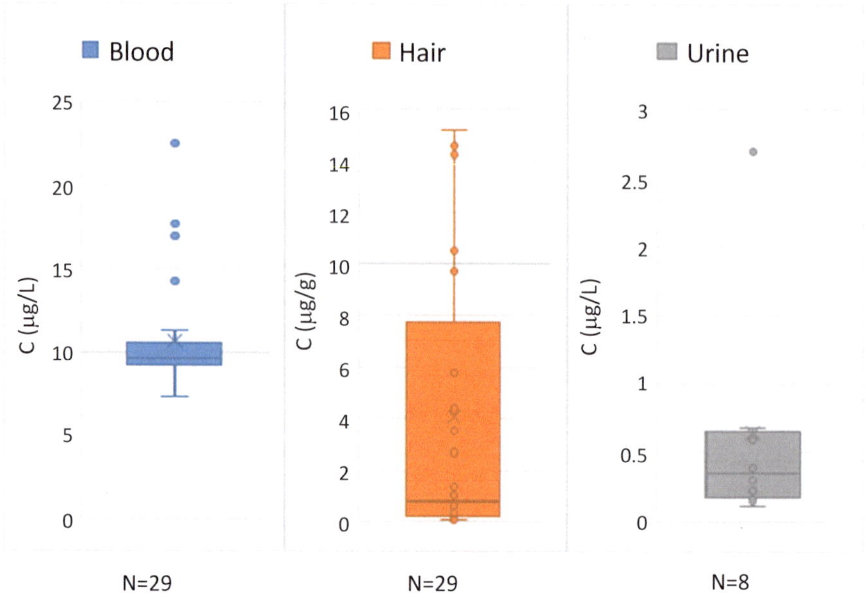

Fig. 4 Box plot of the average Mn levels in biomarkers of exposure reported in the studied literature

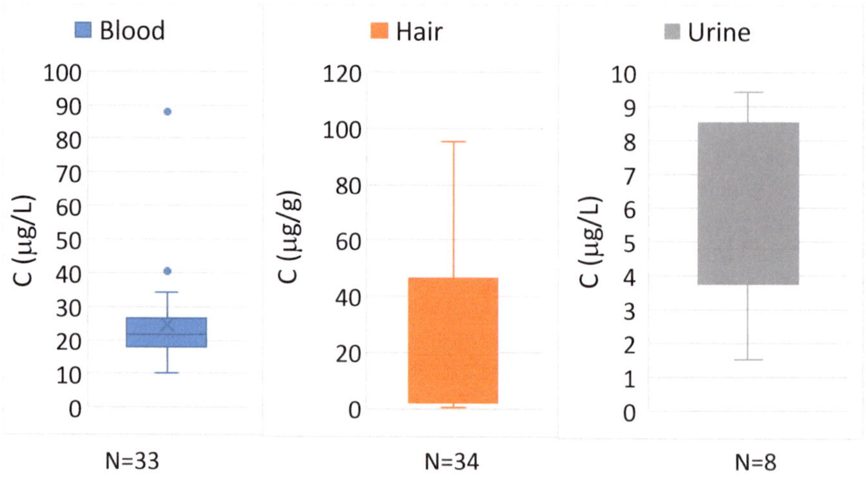

Fig. 5 Box plot of the maximum Mn levels in biomarkers of exposure reported in the studied literature

in the Fourth National Report on Human Exposure to Environmental Chemicals, 9.52 µg/L in 2015/2016 from a sample size of 4,987 participants (US Department of Health and Human Services 2018). Some outliers are observed in Fig. 4 from participants from Korea (Chung et al. 2015; Kim et al. 2009) and Mexico

(De Water et al. 2018; Santos-Burgoa et al. 2001). The narrow distribution of MnB is attributed to (1) a simple and reproducible procedure to measure Mn in blood samples and (2) the poor capacity as a biomarker of chronic low-level environmental exposure to Mn. Higher levels of MnU are observed in Fig. 4 with respect to those found in the Fourth National Report on Human Exposure to Environmental Chemicals: median of 0.35 (this review) vs 0.120 μg/L (2011/2012 from a sample size of 2,504 participants) (US Department of Health and Human Services 2018). MnH distribution is much wider (see Fig. 4), the average ranging from 0.06 to 15.2 μg/g; the lowest value was found in Garda Lake, Italy, a site used as non-exposed area in a study carried out by Eastman et al. (2013) and the highest value in Simoes Filho (Brazil) near a Mn alloy plant. However, the low values of MnH found in other studies close to ferromanganese plants (see, e.g., Eastman et al. (2013) and Haynes et al. (2012)) suggest that these differences should be attributed to different experimental procedures for the determination of Mn in hair samples, mainly in the cleaning process, and probably a fraction of MnH can be considered as exogenous. The potential external contamination of hair might affect the validity to represent internal dose of Mn exposure (Eastman et al. 2013). Figure 5 shows the maximum MnB, MnH, and MnU values reported in the studied literature; it can be observed very high values corresponding to some individuals environmentally exposed to air Mn. For example, Santos-Burgoa et al. (2001) reported a maximum MnB of 88 μg/L in the Hidalgo state (México) near Mn ore mines and Menezes-Filho et al. (2011) a maximum MnH of 95.5 μg/g in Simoes Filho (Brazil) near a ferromanganese plant. However, in the case of MnU, the maximum value (9.4 μg/L) was found in a non-exposed area in Italy (Lago di Garda) (Lucchini et al. 2014).

An appropriate biomarker of Mn exposure should reflect quantitatively exposure-related changes (Zheng et al. 2011); however, this is not always found in the studied literature, mainly for MnB. For example, respirable Mn levels and MnB data measured in participants of a small cross-sectional study in an urban site of Canada carried out by Bolté et al. (2010) suggest that a high atmospheric respirable Mn concentration does not lead to an increase in MnB levels. According to Haynes et al. (2010), MnB measured in adults living near a ferromanganese refinery was not significantly correlated with annual Mn concentration evaluated by AERMOD modeling. In a later work carried out with children, Haynes et al. (2012) observed a borderline relationship between Mn exposure determined by personal samplers and MnB; however, the correlation was not significant. Nevertheless, some studies carried out in a mining district of Mexico (Molango) found positive associations between air Mn and MnB: thus, MnB concentrations were significantly related to the point source in the studied community (Santos-Burgoa et al. 2001); Rodríguez-Agudelo et al. (2006), after adjusting a lineal regression model using logarithmic transformation of Mn values in blood, found an association between MnB and air Mn. In a later work performed in the same area, air and blood Mn levels showed a low but positive correlation ($r = 0.22$, $p < 0.05$) (Solís-Vivanco et al. 2009).

With respect to other biomarkers of exposure, hair and nails seem to better reflect long-term exposure. Menezes-Filho et al. (2009a) reported that the highest MnH levels in children near a ferromanganese plant in Brazil were observed in areas, which correspond to residences closest to this plant, where higher airborne Mn levels are expected to occur. In a later study in the same area, Mn in dust fall accumulation was positively associated with MnH and also with Mn in toenails (Rodrigues et al. 2018b). Lucas et al. (2015) reported a weak association between Mn in dust and saliva in children living near a ferromanganese plant; an association was also found with Mn levels in fingernails. Only Haynes et al. (2010) did not found a significant correlation between MnH and annual Mn concentration modeled at each participant residence in Marietta, Ohio.

3.3 Neurological and Neuropsychological Effects Assessed in Environmentally Airborne Manganese-Exposed Population

Several studies have evaluated different neurological and neuropsychological functions in populations exposed to airborne Mn. The cognitive and motor functions were the most frequently assessed (in 25 and 22 references, respectively); emotional function was evaluated in six references and olfactory function in four references (see Table 4). The results shown below have been structured according to the compendium of neuropsychological tests published by Strauss et al. (2006).

3.3.1 Cognitive Function

Different specific outcomes have been measured in the studied literature to assess the cognitive function. The heterogeneity among the tests to determine cognitive function is higher in adults because different cognitive function domains are also assessed. In children, the cognitive domain is mostly restricted to intelligence.

Intelligence and General Cognitive Functioning

The *Wechsler Adult Intelligence Scale (WAIS)* in its versions III or IV is used in adults. This test provides a full-scale IQ but is also based on subscales or domains. Bowler et al. (2015) provided negative correlations between modeled air Mn and specific domains such as digit span (forward and backward), digit symbol, similarities, or animal naming, reaching in this last domain statistical significance (Spearman correlation coefficient $= -0.19$, $p = 0.01$). Regarding mean differences between exposed and non-exposed groups, Bowler et al. (2012) reported a non-significant negative association in the similarities domain (Bowler et al. 2012) with poorer scores among the exposed adults (Marietta, Ohio) with respect to the comparison town (Mount Vernon, Ohio). In a mining area in Molango (Mexico), Solís-Vivanco et al. (2009) reported a risk of worse punctuations in digit span, with an odds ratio (OR) of 1.75 in the group exposed to air Mn. Viana et al. (2014) and Lucchini et al. (2014) reported as well non-significant negative differences in digit span (forward and backward) in populations near Mn alloy plants in Brazil and Italy, respectively.

The *Neuropsychological Assessment Battery (NAB)* was used also by Bowler et al. (2015) with non-significant negative correlations in its memory immediate and delayed reported domains. Other test used also in adults in the studied literature is the *Mini-Mental State Examination Scale (MMSE)*. Solís-Vivanco et al. (2009) reported a higher risk of worse punctuations in the MMSE among exposed to air Mn with cutoff points at 0.05 and 0.1 μg/m^3 but without yielding statistical significance. Lucchini et al. (2014) reported also similar means among elder population exposed and non-exposed to Mn according to MMSE. These authors also applied the *Raven's Colored Progressive Matrices (CPM)* test, reporting statistically significant negative differences.

The *Wechsler Intelligence Scale for Children (WISC)*, in its versions III or IV, is the most widely applied test to assess the cognitive function of children living in areas exposed to airborne Mn. Overall, a recently published systematic review (Leonhard et al. 2019) concluded that the statistical associations reported in the few studies of specific Mn biomarkers and child intelligence do not allow to establish causal effects based on the Bradford Hill considerations, highlighting the methodological problems associated with cross-sectional designs or the lack of confusion control in some studies without providing adjusted estimations. However, negative significant associations between airborne Mn exposure and child full-scale IQ and/or subscale scores were found in different areas near Mn sources in approaches based on adjusted b coefficients (difference of means) or correlation coefficients: e.g., close to Mn alloy plants in Simoes Filho, Brazil (Carvalho et al. 2014, 2018; Menezes-Filho et al. 2011), Ohio, USA (Haynes et al. 2015, 2018), four areas from Korea (Kim et al. 2009), or Molango, Mexico, near Mn ore mines (Riojas-Rodríguez et al. 2010).

Interestingly, Chung et al. (2015) have reported nonlinear inverted U-shaped associations between maternal MnB concentration and 6-month Mental Development Index (MDI) scores as determined by the *Bayley Scales of Infant Development-II (BSID-II)*, by using penalized spline regression. Based on the smoothed models with the smallest Akaike's information criterion (AIC) values, associations with increasing concentrations of maternal MnB changed from positive to negative at 24–26 μg/L MnB. The same pattern was observed also for motor function according to Psychomotor Development Index (PDI) also determined by the BSID-II.

Attention and Cognitive Flexibility

In adults, the attention and visual processing speed are commonly assessed by the *Trail Making Test (TMT)* part A, whereas TMT part B evaluates divided attention and cognitive flexibility. Negative significant differences have been reported according to this test in approaches based on adjusted b coefficients, in elder population living near closed Mn alloy plants in Valcamonica, Italy, in relation to MnU (Lucchini et al. 2014) and near an active ferromanganese plant in Brazil according to axillary hair (Viana et al. 2014); a negative correlation (Spearman correlation coefficient) between modeled air Mn and TMT part B yielded statistical significance near a Mn alloy plant in Marietta, Ohio (Bowler et al. 2015).

Non-statistically significant mean differences or correlations have been reported on other tests evaluating attention such as the *continuous performance test* (Mergler et al. 1999), the *visual attention computerized test-3 (TAVIS-3)* (Viana et al. 2014), or the *auditory consonant trigrams (ACT)* (Bowler et al. 2015) in adults. Lastly, Solís-Vivanco et al. (2009) reported a non-significant higher risk of worse scores in the *word association test* (phonological fluency) (OR = 1.32).

In children, non-significant negative differences were found according to TAVIS-3 in Brazil (Carvalho et al. 2014). A worse performance in visuospatial and attention tasks was reported in Tarragona (Spain) by children living in industrialized areas, according to the *Caras test* (Torrente et al. 2005).

Executive Functions

In adults, a negative correlation of borderline non-significance (Spearman correlation coefficient = −0.16, $p = 0.06$) has been reported on this specific domain as assessed by the *stroop color-word test* (Bowler et al. 2015). However, according to the *clock test*, exposed to air Mn did not show a higher risk of worse punctuations (OR = 0.99) (Solís-Vivanco et al. 2009).

In children, executive functions have been evaluated according to the *Wisconsin Card Sorting Test (WCST)* by Carvalho et al. (2014) in Brazil. After dividing by age groups, non-significant correlations regarding MnH were reported in children aged 7–8 years and 9–10 years, whereas a significant negative correlation was reported in children 11–12 years old (rho = −0.51; p = 0.02). Non-significant negative differences were reported with respect to this domain as assessed by the *inhibition test* (Carvalho et al. 2018).

Memory and Fluency

Statistically significant negative differences (Mergler et al. 1999) or correlations (Bowler et al. 2015) have been reported for this domain as assessed by the *Rey-Osterrieth Complex Figure (ROCF) test*. According to the *Rey Auditory Verbal Learning Test (RAVLT)*, Viana et al. (2014) reported a borderline non-significant difference of medians between groups living at different distances from a source of air Mn: a median of 12 in Cotegipe (1.5 km from the source) in comparison with a median of 14 in Santa Luzia (2.5 km from the source). In contrast, Lucchini et al. (2014) in Italy did not found a mean difference between exposed and non-exposed according to the *Story Recall Test*. A slightly higher risk (non-significant) of worse punctuations (OR = 1.17) has been reported by Solís-Vivanco et al. (2009) for this domain according to the *Word List Test*, whereas no associations were found in the same study in relation to the *semi-complex design test* (OR visual memory coding, 0.99, and OR visual memory recall, 1.05) (Solís-Vivanco et al. 2009).

In children, for the ROCF, no significant differences between groups were reported by Hernández-Bonilla et al. (2016) in Mexico, for the total scores of ROCF copy and ROCF immediate recall. However, in specific analysis for the ROCF copy trial, MnH was significantly associated with an increase in distortion errors (tangency, closure), angle errors, and overtracing (partial overtracing). In the immediate recall trial, MnH was significantly associated with increased overtracing (partial overtracing) and omissions and negatively associated with the number of perceptual drawn units, total score, and percentage immediate recall. In relation to the *Corsi Blocks (forward)*, Carvalho et al. (2014) in Brazil have found a −0.5 points in the children difference of medians against the highest tertil of exposure to Mn ($p = 0.048$). Statistically significant lower scores for all the *Children's Auditory Verbal Learning Test (CAVLT)* subscales were reported in Molango (Mexico) by Torres-Agustín et al. (2013). Verbal memory and fluency have been also determined by using the *List Memory (LM) test* and *Word Generation (WG) test*, respectively, showing non-statistically significant negative differences (against Mn exposed) (Carvalho et al. 2018).

3.3.2 Motor Function

It is well-known that motor function is affected by the occupational exposure to airborne Mn. Some welders and workers from the Mn alloy industry suffer from manganism, a Parkinson-like disease. An increased prevalence of Parkinsonian disturbances has also been detected in non-occupational exposure near former ferromanganese plants that were in operation for decades in Valcamonica, Italy, where high Mn levels in settled dust were found (Lucchini et al. 2007). These authors identified Parkinsonian disorders by means of clinical registers of Parkinson-like diseases and computerized L-Dopa prescription records from the local databank of the Italian health system.

In a more specific way, as for cognitive function, environmental epidemiological studies have tried to evaluate the motor function, through different domains evaluated with a battery of heterogeneous tests that are summarized below.

Motor Speed

Statistically significant negative differences (Lucchini et al. 2014) or correlations (Spearman correlation coefficient = −0.26, $p = <0.001$ with dominant hand; and −0.16, $p = 0.03$ with non-dominant hand) (Bowler et al. 2016) have been reported for this domain as assessed by the *finger-tapping test (FTT)* (dominant and non-dominant hands). Bowler et al. (2012) found however similar means in FTT scores between exposed and non-exposed groups according to MnB.

Beuter et al. (1999) have also reported statistically significant negative differences in relation to the maximum slope domain determined by *Diadochokinesimeter (DIADO)*. For the rest of dimensions determined by DIADO, non-significant negative differences were found.

Regarding children, statistically significant negative differences have been reported by Hernández-Bonilla et al. (2016) in a mining area of Mexico as assessed by FTT.

Motor Coordination and Postural Balance

In adults, statistically significant negative differences have been reported for this domain as assessed by *Luria Nebraska Neuropsychological Battery (LNNB)* (Lucchini et al. 2014) and *postural balance testing* (sway area and length) (Standridge et al. 2008) near ferromanganese plants in Italy and Ohio, USA, respectively. Eye-hand coordination and proximal/distal pointing were measured by *Eurythmokinesimeter (EKM)* by Beuter et al. (1999), reporting statistically significant negative differences in adults >50 years. A higher risk of worse punctuations in the *Ostrosky-Solís's neuropsychological battery* has been reported by Rodríguez-Agudelo et al. (2006), except in the reproduce hand positions domains. Lucchini et al. (2014) have reported lastly non-significant results according to the *body sway assessment test*.

In children, statistically significant negative differences have been reported in relation to the LNNB (Lucchini et al. 2012; Chiu et al. 2017) and the *Hall-Effect type portable force platform system* (Rugless et al. 2014).

Tremor

In adults, arm and hand tremor was evaluated by Beuter et al. (1999) by a *tremor system*, reporting statistically significant negative differences in the frequency dispersion subdomain and non-significant differences in the tremor intensity, center frequency, and harmonic index subdomains. Statistically significant negative differences have been later reported for tremor as assessed by *CATSYS tremor system* (Bowler et al. 2016; Kim et al. 2011). Lastly, no differences in some subdomains (harmonic index, right and left) or non-statistically significant negative differences in other subdomains (intensity, right and left) have been reported by Lucchini et al. (2014) according to their applied *tremor assessment*.

In children, statistically significant negative differences have been reported by Lucchini et al. (2012), in relation to the *Tremor 7.0 of Danish Products Developments-DPD test*. Resting tremor was assessed using the *Tremor Pen Test* in the CATSYS tremor system by Chiu et al. (2017) in Italy, reporting that higher early postnatal Mn exposure was associated with increased right-hand center frequency in girls (p for interaction <0.01) and that an increased Mn level at the later postnatal period was associated with increased center frequency in boys (p for interaction $= 0.01$).

Bradykinesia

The *Unified Parkinson's Disease Rating Scale (UPDRS)*, Part II (motor-related activities of daily living (ADL) and Part III (motor/movement examination)), has been applied to population living near air Mn sources with statistically significant negative differences in motor and bradykinesia scores (Bowler et al. 2012; Mergler et al. 1999). Kim et al. (2011) have reported a higher risk of worse punctuations in the bradykinesia (OR $= 2.9$) and motor scores (OR $= 2.4$).

Hand Dexterity

In Canada statistically significant negative differences were reported in relation to the *Purdue Pegboard Test* in adults >50 years (Mergler et al. 1999). Later, statistically significant negative differences (Bowler et al. 2012) or negative correlations (Spearman correlation coefficient $= -0.10$, $p = 0.17$ with dominant hand; and -0.15, $p = 0.04$ with non-dominant hand) (Bowler et al. 2016) have been reported for this domain as assessed by the *Grooved Pegboard Test (GPT)*. In contrast, no differences were found with this test in Brazil (Viana et al. 2014).

The hand *dynamometer* has been used to measure strength manipulative function in Bowler et al. (2012, 2016) studies, providing significant negative differences (Bowler et al. 2012) and non-statistically significant negative correlations (Spearman correlation coefficient $= -0.06$, $p = 0.4$ with dominant hand; and -0.05, $p = 0.47$ with non-dominant hand) (Bowler et al. 2016).

In children, statistically significant negative differences have been reported with the *Grooved Pegboard Task* near a ferromanganese plant in Brazil (Carvalho et al. 2018). The *Pursuit Aiming Test* has been used also in children and adolescents. Statistically significant negative differences have been reported in one study (Lucchini et al. 2012), and sex differences in sensitivity to prenatal and early childhood Mn exposure on neuromotor function in adolescents were compared in another study reporting that higher exposure was associated with more error dots in girls, compared to boys, although none of the interaction models were statistically significant (Chiu et al. 2017).

3.3.3 Olfactory Function

The olfactory function has been assessed by the *Sniffin' Sticks test* in adults, with statistically significant negative differences being reported (Guarneros et al. 2013; Lucchini et al. 2014; Mergler et al. 1999).

In children, statistically significant negative differences have also been reported in relation to the *Sniffin' Sticks task* (Lucchini et al. 2012).

3.3.4 Emotional Function and Mood

The emotional function has been determined in Mn-exposed population by assessing the anxiety using the *Symptom Checklist-90 Revised (SCL-90-R)*. Statistically significant negative differences have been reported by Bowler et al. (2012).

Depression has been assessed by the *Geriatric Depression Inventory (GDI)* and the *Beck Depression Inventory (BDI)*. Solís-Vivanco et al. (2009) have reported non-significant higher risks of worse punctuations in GDI in exposed adults according to the cutoff points of 0.05 and 0.1 $\mu g/m^3$. However, for the BDI the positive association (OR = 3.01; 95% CI 0.66–13.65) became the opposite (OR = 0.55; 95% CI 0.20–1.51) when the cutoff point increased from 0.05 to 0.1 $\mu g/m^3$.

Regarding children, externalizing (disruptive and aggressive) and internalizing behaviors (withdrawn, somatic complaints, and anxious/depressed scales) have been assessed by using the *Conner's Abbreviated Teacher Rating Scale*. Carvalho et al. (2018) have reported statistically significant negative differences in Brazil according to this scale. Other studies have used the *Child Behavior Checklist (CBCL)*, reporting also statistically significant negative differences with poorer scores among the exposed children (Menezes-Filho et al. 2014; Rodrigues et al. 2018a).

4 Discussion and Recommendations

A growing concern about the potential negative neurological/neuropsychological effects of environmental exposure to Mn is shown in recent literature. Among the exposure routes, inhalation of airborne Mn is the most important one in areas located in the vicinity of Mn industrial sources, such as Mn alloy facilities, Mn ore mines, and plants where these ores are processed. This literature review shows firstly the wide distribution of Mn levels in PM matrices (TSP, PM10, PM2.5, respirable); in the studies performed near industrial sources of air Mn, these levels usually exceed the US EPA RfC, i.e., 50 ng/m^3, (87.5% in TSP, 47% in PM10 and 37.5% in PM2.5) and in some cases the WHO (150 ng/m^3) and ATSDR (300 ng/m^3) annual guidelines (75/37.5%, 40/27%, and 12.5/0% in TSP, PM10, and PM2.5, respectively). Although some controversial studies have been published about the necessity of increasing the acceptable Mn ambient air concentration (Bailey et al. 2009; Gentry et al. 2017), according to the precautionary principle, it seems reasonable to maintain at least the WHO annual guideline until new evidences allow to change it. In the meantime, it is also important that regions without limit/threshold/guideline values for air Mn, such as the European Union, do establish them in new regulations.

Numerous attempts have been made to select an appropriate biomarker of Mn chronic environmental exposure; however, at the moment there is not a unique valid biomarker of exposure (Coetzee et al. 2016; Leonhard et al. 2019). Although MnH has been proposed by some authors as a consistent biomarker of chronic long-term exposure to Mn (Bouchard et al. 2011; Coetzee et al. 2016; Eastman et al. 2013), the wide distribution found in the boxplot shown in Fig. 4 makes unfeasible the comparison between different populations exposed to airborne Mn. For example, near active ferromanganese plants, a MnH average value of 14.6 μg/g was reported by Carvalho et al. (2014) in Simoes Filho, Brazil, while Eastman et al. (2013) reported an average value two order of magnitude lower (0.134 μg/g) near Bagnolo Mella, Italy. In this case, it seems that the different cleaning, processing, and analysis protocols and external contamination of hair near industrial sources of Mn may cause such huge differences (Coetzee et al. 2016; Haynes et al. 2015). This also may explain the differences in the correlation between environmental exposure and MnH found in some studies (Haynes et al. 2010; Lucas et al. 2015; Rodrigues et al. 2018b). With respect to MnB, it may describe changes in the short-term Mn exposure (Bouchard et al. 2011); however, due to the homeostatic control of Mn levels in the human body, a short half-time of Mn in blood with respect to tissue and intracellular compartments is assumed (Gunier et al. 2014; Zheng et al. 2011), and therefore MnB is usually considered as a poor biomarker of Mn exposure, at least for chronic exposure. In any case, the range of MnB levels is very narrow with the exception of some outliers (see Fig. 4); three of the four outliers shown in this figure are reported from studies where no airborne Mn levels were measured, so they cannot be attributed to the environmental exposure to high air Mn levels. These high average MnB levels are explained by very high concentrations found in some participants (e.g., up to 88 μg/L in Santos-Burgoa et al. 2001).

The low growing speed of nails, about 6 months before collection for fingernails and 7–12 months for toenails (Rodrigues et al. 2018b) makes them a promising biomarker of Mn chronic exposure (Lucas et al. 2015; Rodrigues et al. 2018b); however, at the moment, a few studies have used nails as biomarkers (Lucas et al. 2015; Reis et al. 2015; Rodrigues et al. 2018a, 2018b; Viana et al. 2014), and therefore further studies that investigate positive associations between environmental airborne Mn exposure and Mn levels in toe and fingernails are needed to consider nails as a valid biomarker. Other promising biomarker of Mn exposure is dentine; the measurement of Mn levels in dentine of shed teeth represents prenatal, early postnatal, and cumulative childhood exposure to Mn (Chiu et al. 2017; Gunier et al. 2014, 2015; Mora et al. 2015). However, this approach can only be applied to children, and the analytical procedure presents a high complexity.

Independently of the type of biomarker used, there is a need to check the correlation between environmental sources and biomarkers (Coetzee et al. 2016). The traditional approach estimates the environmental airborne Mn exposure by the measurement or modeling of the total concentration of Mn in a PM fraction (TSP, PM10, PM2.5, respirable); however, it has been suggested recently that the bioaccessible concentration may better reflect the risk of exposure to a given pollutant (Hernández-Pellón et al. 2018; Huang et al. 2016, 2018). Ellingsen et al.

(2013) studied for the first time the relationship between the bioaccessibility of Mn in welding fume particles and blood and urine concentrations of welders. Good correlations between Mn concentrations in a lung fluid (Hatch solution) after contacting it with welding fumes and the concentration of Mn in biological fluids were found. So, they suggested that the study of the solubility of Mn in lung fluids might contribute to understand the conflicting results in biomonitoring studies related to Mn exposure. This approach may be applied to environmental studies, i.e., the measurement of the bioaccessible content of Mn in PM matrices collected with personal samplers and the association between the bioaccessible concentration and selected biomarkers of Mn exposure.

This literature review has shown that cognitive and motor functions are frequently assessed in air Mn-exposed population (25 and 22 references, respectively); in some studies, mood and the emotional and olfactory functions are also evaluated. An overview of the main results and conclusions derived from the studied literature shows that some neurological/neuropsychological impairments, mainly motor, are associated with Mn exposure predominantly measured by MnH, but not always this association is statistically significant. In addition, the design of these studies is heterogeneous: different environmental exposure to airborne Mn measured in various PM matrices, as shown in Fig. 2, the use of different biomarkers of exposure, different pre-treatment and analysis protocols of biological samples, different tests and parts of these tests to assess the same neurological/neuropsychological outcomes, different data analysis strategies, and the use of different potential confounders in statistical models. So, the different characteristics of these studies make difficult the comparison between the conclusions obtained from these works.

From an epidemiological point of view, the main confounding variables would be age, ethnicity/race, sex, and education (educational level). The latter would also act as a surrogate for socioeconomic level or social class. It is important that results in tests (crude scores) are standardized for these variables whenever possible (Heaton et al. 2004).

Regarding the tests used to evaluate cognitive and motor functions, mood, and emotional and olfactory functions, they should have proven reliability and validity, especially in terms of sensitivity. It is important to note that we study the health effects of environmental but non-occupational exposures, which implies lower levels of exposure. With respect to cognitive function, the different tests reported by studies commonly used in the world and many countries have norms for their standardization with respect to population values by age and sex. With respect to motor function, computerized neuromotor tests such as those described by Beuter et al. (1999), or the CATSYS system to evaluate tremor and body sway (Després et al. 2000), seem sufficiently sensitive, specific, and reliable.

Statistical approaches to primary studies should include at least (1) a correlation approach (the higher the Mn levels, the poorer scores) and (2) a mean differences approach, comparing the mean scores in tests between exposed and non-exposed. This approach allows the results of different tests to be combined into standardized mean differences (SMDs) as a summary measure of effect in meta-analysis. As the SMD is standardized, the pooled "effect size" obtained through a meta-analysis makes it possible to interpret the magnitude of the "effect size," according to the criteria of Cohen (1988).

The most important question to solve here is if it is possible to demonstrate evidence of a causality in the association between environmental exposure to airborne Mn and neurological/neuropsychological outcomes in children and adults. The majority of those studies are cross-sectional in design; according to Leonhard et al. (2019), the temporal sequence between Mn exposure and neurodevelopmental outcomes (child intelligence) cannot be resolved in such epidemiological studies, and only prospective cohort studies might demonstrate causality between exposure and effects. However, if biomarkers that account for long-term Mn exposure periods (months or years) or biomarkers that clearly correlate well with environmental airborne Mn levels are used, a cause-effect relationship can be considered.

Lastly, the clinical and public health impact remains to be defined. This would require studies specifically developed according to these objectives.

5 Conclusions

The literature review performed in this work shows the growing concern about the potential negative neurological/neuropsychological effects of environmental exposure to airborne Mn; overall, among 47 articles that matched the search criteria, 31 studies reported neurological/neuropsychological outcomes that were assessed in the Mn-exposed population.

The levels of Mn in air differ among the studied references depending on the type of source and distance from it, which makes difficult the comparison between those studies. In addition, different evaluation methods of Mn levels, i.e., modeling and measurement with stationary and personal PM samplers, are used. Anyway, the airborne Mn average levels close to Mn industrial sources usually exceed the US EPA RfC (i.e., 50 ng/m^3) and in some cases the WHO and ATSDR annual guidelines (150 and 300 ng/m^3, respectively). In all of these studies, total Mn concentration in a PM fraction was considered instead of the bioaccessible concentration, which may reflect better the exposure to airborne Mn. None of the studied biomarkers seem to reflect properly changes in the exposure to environmental airborne Mn. Among the biomarkers of exposure used in the literature, MnH has been considered to be the most promising; however, poor associations with levels of Mn in environmental airborne PM and a wide distribution of MnH values among the studies, probably due to the external contamination of hair samples and different cleaning and analysis protocols, suggest that further studies should be done to confirm the adequacy of MnH as biomarker of long-term chronic environmental exposure to airborne Mn. Other biomarkers, such as blood and urine, show smaller differences between studies not attributable to analytical methods, but they do not seem to reflect chronic exposure to Mn. Toe and fingernails have been used in a few recent articles, but the small number of studies does not allow us to consider nails as a valid biomarker at the moment. So, the first important gap to fill in is to find an appropriate biomarker of exposure that reflects the environmental chronic exposure to sources of air Mn; the measurement of the bioaccessible content of Mn in PM matrices collected with personal samplers may help in finding the association between exposure to airborne sources of Mn and biomarkers.

Among the neurological/neuropsychological outcomes studied in air Mn-exposed population, the cognitive and motor functions were the most frequently assessed (25 and 22 references, respectively); only a few references evaluated mood and the emotional and olfactory functions. Although most of those studies are cross-sectional in design and therefore causality between exposure to Mn and neurological/neuropsychological effects is difficult to demonstrate, they show that some neurological/neuropsychological outcomes, mainly motor, are negatively significantly associated with Mn exposure, mainly measured by MnH. However, the heterogeneity of those studies, i.e., different levels of air Mn, biomarkers of exposure and analytical procedures to determine them, neurological/neuropsychological tests, statistical approaches, etc., makes difficult the comparison between the results and conclusions derived from them.

In conclusion, further studies with a most systematic methodology for estimating the environmental airborne Mn exposure, for example, through the measurement of the bioaccessible content of airborne Mn with personal samplers, using a proper biomarker of chronic exposure and using a standardized homogeneous battery of neurological/neuropsychological tests, are needed to clarify the effects of chronic, low-/medium-dose environmental exposure to airborne Mn.

Acknowledgments This work was supported by the Spanish Ministry of Science, Innovation and Universities through the CTM2017-82636-R Project. This funding source was not involved in the study design; data collection, analysis, or interpretation; the writing of the article; or the decision to submit for publication.

Conflict of Interest The authors declare that they have no conflict of interest.

References

Aschner M, Guilarte TR, Schneider JS, Zheng W (2007) Manganese: recent advances in understanding its transport and neurotoxicity. Toxicol Appl Pharmacol 221(2):131–147

ATSDR (2012) Toxicological profile for manganese. U.S. Department of Health and Human Services, Public Health Service, Atlanta

Bader M, Dietz MC, Ihrig A, Triebig G (1999) Biomonitoring of manganese in blood, urine and axillary hair following low-dose exposure during the manufacture of dry cell batteries. Int Arch Occup Environ Health 72:521–527

Bailey LA, Goodman JE, Beck BD (2009) Proposal for a revised reference concentration (RfC) for manganese based on recent epidemiological studies. Regul Toxicol Pharmacol 55(3):330–339

Baldwin M, Mergler D, Larribe F, Belanger S, Tardif R, Bilodeau L, Hudnell K (1999) Bioindicator and exposure data for a population based study of manganese. Neurotoxicology 20 (2–3):343–354

Batterman S, Su F-C, Jia C, Naidoo RN, Robins T, Naik I (2011) Manganese and lead in children's blood and airborne particulate matter in Durban, South Africa. Sci Total Environ 409 (6):1058–1068

Beuter A, Edwards R, de Geoffroy A, Mergler D, Hundnell K (1999) Quantification of neuromotor function for detection of the effects of manganese. Neurotoxicology 20:355–366

Bjørklund G, Chartrand MS, Aaseth J (2017) Manganese exposure and neurotoxic effects in children. Environ Res 155:380–384

Bolté S, Normandin L, Kennedy G, Zayed J (2010) Human exposure to respirable manganese in outdoor and indoor air in urban and rural areas. J Toxicol Environ Health A 67(6):459–467

Bouchard MF, Sauvé S, Barbeau B, Legrand M, Brodeur MÈ, Bouffard T, Limoges E, Bellinger DC, Mergler D (2011) Intellectual impairment in school-age children exposed to manganese from drinking water. Environ Health Perspect 119(1):138–143

Bowler RM, Gysens E, Diamond S, Nakagawa M, Drezgic M, Roels HA (2006a) Manganese exposure: neuropsychological and neurological symptoms and effects in welders. Neurotoxicology 27:315–326

Bowler RM, Koller W, Schulz PE (2006b) Parkinsonism due to manganism in a welder: neurological and neuropsychological sequelae. Neurotoxicology 27:237–332

Bowler RM, Harris M, Gocheva V, Wilson K, Kim Y, Davis SI, Bollweg G, Lobdell DT, Ngo L, Roels HA (2012) Anxiety affecting parkinsonian outcome and motor efficiency in adults of an Ohio community with environmental airborne manganese exposure. Int J Hyg Environ Health 215(3):393–405

Bowler RM, Kornblith ES, Gocheva VV, Colledge MA, Bollweg G, Kim Y, Beseler CL, Wright CW, Adams SW, Lobdell DT (2015) Environmental exposure to manganese in air: associations with cognitive functions. Neurotoxicology 49:139–148

Bowler RM, Beseler CL, Gocheva VV, Colledge M, Kornblith ES, Julian JR, Kim Y, Bollweg G, Lobdell DT (2016) Environmental exposure to manganese in air: associations with tremor and motor function. Sci Total Environ 541:646–654

Carter MR, Gaudet BJ, Stauffer DR, White TS, Brantley SL (2015) Using soil records with atmospheric dispersion modeling to investigate the effects of clean air regulations on 60 years of manganese deposition in Marietta, Ohio (USA). Sci Total Environ 515–516:49–59

Carvalho CF, Menezes-Filho JA, Matos VPD, Bessa JR, Coelho-Santos J, Viana GFS, Argollo N, Abreu N (2014) Elevated airborne manganese and low executive function in school-aged children in Brazil. Neurotoxicology 45:301–308

Carvalho CFD, Oulhote Y, Martorelli M, Carvalho COD, Menezes-Filho JA, Argollo N, Abreu N (2018) Environmental manganese exposure and associations with memory, executive functions, and hyperactivity in Brazilian children. Neurotoxicology 69:253–259

Chen P, Culbreth M, Aschner M (2016) Exposure, epidemiology, and mechanism of the environmental toxicant manganese. Environ Sci Pollut Res 23:13802–13810

Chen P, Bornhorst J, Aschner M (2018) Manganese metabolism in humans. Front Biosci 23:1655–1679

Chiu YH, Henn B, Hsu HH, Pendo M, Coull B, Austin C, Cagna G, Fedrighi C, Placidi D, Smith D, Wright R, Lucchini R, Arora M (2017) Sex differences in sensitivity to prenatal and early childhood manganese exposure on neuromotor function in adolescents. Environ Res 159:458–465

Chung SE, Cheong H-K, Ha E-H, Kim B-N, Ha M, Kim Y, Hong Y-C, Park H, Oh S-Y (2015) Maternal blood manganese and early neurodevelopment: the mothers and children's environmental health (MOCEH) study. Environ Health Perspect 123(7):717–722

Coetzee DJ, McGovern PM, Rao R, Harnack LJ, Georgieff MK, Stepanov I (2016) Measuring the impact of manganese exposure on children's neurodevelopment: advances and research gaps in biomarker based approaches. Environ Health 15:91

Cohen J (1988) Statistical power analysis for the behavioural sciences, 2nd edn. Routledge, New York

De Water E, Proal E, Wang V, Medina SM, Schnaas L, Téllez-Rojo MM, Wright RO, Tang CY, Horton MK (2018) Prenatal manganese exposure and intrinsic functional connectivity of emotional brain areas in children. Neurotoxicology 64:85–93

Després C, Lamoureux D, Beuter A (2000) Standardization of a neuromotor test battery: the CATSYS system. Neurotoxicology 21:1–11

Eastman RR, Jursa TP, Benedetti C, Lucchini RG, Smith DR (2013) Hair as a biomarker of environmental manganese exposure. Environ Sci Technol 47(3):1629–1637

Ellingsen DG, Hetland SM, Thomassen Y (2003) Manganese air exposure assessment and biological monitoring in the manganese alloy production industry. J Environ Monit 5:84–90

Ellingsen DG, Zibarev E, Kusraeva Z, Berlinger B, Chashchin M, Bast-Pettersen R, Chashchin V, Thomassen Y (2013) The bioavailability of manganese in welders in relation to its solubility in welding fumes. Environ Sci Processes Impacts 15:357–365

Erikson KM, Thompson K, Aschner J, Aschner M (2007) Manganese neurotoxicity: a focus on the neonate. Pharmacol Ther 113(2):369–377

Gentry PR, Van Landingham C, Fuller WG, Sulsky SI, Greene TB et al (2017) A tissue dose-based comparative exposure assessment of manganese using physiologically based pharmacokinetic modeling- the importance of homeostatic control for an essential metal. Toxicol Appl Pharmacol 322:27–40

Guarneros M, Ortiz-Romo N, Alcaraz-Zubeldia M, Drucker-Colín R, Hudson R (2013) Nonoccupational environmental exposure to manganese is linked to deficits in peripheral and central olfactory function. Chem Senses 38(9):783–791

Gunier RB, Mora AM, Smith D, Arora M, Austin C, Eskenazi B, Bradman A (2014) Biomarkers of manganese exposure in pregnant women and children living in an agricultural community in California. Environ Sci Technol 48(24):14695–14702

Gunier RB, Arora M, Jerrett M, Bradman A, Harley KG, Mora AM, Kogut K, Hubbard A, Austin C, Holland N, Eskenazi B (2015) Manganese in teeth and neurodevelopment in young Mexican-American children. Environ Res 142:688–695

Haynes EN, Heckel P, Ryan P, Roda S, Leung Y-K, Sebastian K, Succop P (2010) Environmental manganese exposure in residents living near a ferromanganese refinery in Southeast Ohio: a pilot study. Neurotoxicology 31(5):468–474

Haynes EN, Ryan P, Chen A, Brown D, Roda S, Kuhnell P, Wittberg D, Terrell M, Reponen T (2012) Assessment of personal exposure to manganese in children living near a ferromanganese refinery. Sci Total Environ 427-428:19–25

Haynes EN, Sucharew H, Kuhnell P, Alden J, Barnas M, Wright RO, Parsons PJ, Aldous KM, Praamsma ML, Beidler C, Dietrich KN (2015) Manganese exposure and neurocognitive outcomes in rural school-age children: the communities actively researching exposure study (Ohio, USA). Environ Health Perspect 123(10):1066–1071

Haynes EN, Sucharew H, Hilbert TJ, Kuhnell P, Spencer A, Newman NC, Burns R, Wright R, Parsons PJ, Dietrich KN (2018) Impact of air manganese on child neurodevelopment in East Liverpool, Ohio. Neurotoxicology 64:94–102

He P, Liu DH, Zhang GQ (1994) Effects of high-level-manganese sewage irrigation on children's neurobehavior. Chin J Prev Med 28(4):216–218

Heaton RK, Miller WS, Taylor MJ, Grant I (2004) Revised comprehensive norms for an expanded Halstead-Reitan battery: demographically adjusted neuropsychological norms for African American and Caucasian adults. Psychol Assess Resour

Hernández-Bonilla D, Escamilla-Núñez C, Mergler D, Rodríguez-Dozal S, Cortez-Lugo M, Montes S, Tristán-López LA, Catalán-Vázquez M, Schilmann A, Riojas-Rodriguez H (2016) Effects of manganese exposure on visuoperception and visual memory in schoolchildren. Neurotoxicology 57:230–240

Hernández-Pellón A, Nischkauer W, Limbeck A, Fernández-Olmo I (2018) Metal(loid) bioaccessibility and inhalation risk assessment: A comparison between an urban and an industrial area. Environ Res 165:140–149

Huang X, Betha R, Tan LY, Balasubramanian R (2016) Risk assessment of bioaccessible trace elements in smoke haze aerosols versus urban aerosols using simulated lung fluids. Atmos Environ 125:505–511
Huang H, Jiang Y, Xu X, Cao X (2018) In vitro bioaccessibility and health risk assessment of heavy metals in atmospheric particulate matters from three different functional areas of Shanghai. Chin Sci Total Environ 610–611:546–554
Hudnell HK (1999) Effects from environmental manganese exposure: A review of the evidence from non-occupational studies. Neurotoxicology 20:379–398
Khan K, Wasserman GA, Liu X, Ahmed E, Parvez F, Slavkovich V, Levy D, Mey J, van Geen A, Graziano JH, Factor-Litvak P (2012) Manganese exposure from drinking water and children's academic achievement. Neurotoxicology 33(1):91–97
Kim Y, Kim BN, Hong YC, Shin MS, Yoo HJ, Kim JW, Bhang SY, Cho SC (2009) Co-exposure to environmental lead and manganese affects the intelligence of school-aged children. Neurotoxicology 30:564–571
Kim Y, Bowler RM, Abdelouahab N, Harris M, Gocheva V, Roels HA (2011) Motor function in adults of an Ohio community with environmental manganese exposure. Neurotoxicology 32 (5):606–614
Kornblith ES, Casey SL, Lobdell DT, Colledge MA, Bowler RM (2018) Environmental exposure to manganese in air: tremor, motor and cognitive symptom profiles. Neurotoxicology 64:152–158
Leonhard MJ, Chang ET, Loccisano AE, Garry MR (2019) A systematic literature review of epidemiologic studies of developmental manganese exposure and neurodevelopmental outcomes. Toxicology 420:46–65
Lin Y-Y, Leon Guo Y-L, Chen P-C, Liu J-H, Wu H-C, Hwang Y-H (2011) Associations between petrol-station density and manganese and lead in the cord blood of newborns living in Taiwan. Environ Res 111(2):260–265
Lucas EL, Bertrand P, Guazzetti S, Donna F, Peli M, Jursa TP, Lucchini R, Smith DR (2015) Impact of ferromanganese alloy plants on household dust manganese levels: implications for childhood exposure. Environ Res 138:279–290
Lucchini R, Apostoli P, Perrone C, Placidi D, Albini E, Migliorati P, Mergler D, Sassine MP, Palmi S, Alession L (1999) Long-term exposure to "low levels" of manganese oxides and neurofunctional changes in ferroalloy workers. Neurotoxicology 20:287–298
Lucchini RG, Albini E, Benedetti L, Borghesi S, Coccaglio R, Malara EC, Parrinello G, Garattini S, Resola S, Alessio L (2007) High prevalence of Parkinsonian disorders associated to manganese exposure in the vicinities of ferroalloy industries. Am J Ind Med 50(11):788–800
Lucchini RG, Guazzetti S, Zoni S, Donna F, Peter S, Zacco A, Salmistraro M, Bontempi E, Zimmerman NJ, Smith DR (2012) Tremor, olfactory and motor changes in Italian adolescents exposed to historical ferro-manganese emission. Neurotoxicology 33(4):687–696
Lucchini RG, Guazzetti S, Zoni S, Benedetti C, Fedrighi C, Peli M, Donna F, Bontempi E, Borgese L, Micheletti S, Ferri R, Marchetti S, Smith DR (2014) Neurofunctional dopaminergic impairment in elderly after lifetime exposure to manganese. Neurotoxicology 45:309–317
Lucchini RG, Aschner M, Landrigan PJ, Cranmer JM (2018) Neurotoxicity of manganese: indications for future research and public health intervention from the manganese 2016 conference. Neurotoxicology 64(1):1–4
Mattison DR, Milton B, Krewski D, Levy L, Dorman DC, Aggett PJ, Roels HA, Andersen ME, Karyakina NA, Shilnikova N, Ramoju S, McGough D (2017) Severity scoring of manganese health effects for categorical regression. Neurotoxicology 58:203–216
Menezes-Filho JA, Paes CR, de Pontes AMC, Moreira JC, Sarcinelli PN, Mergler D (2009a) High levels of hair manganese in children living in the vicinity of a ferro-manganese alloy production plant. Neurotoxicology 30(6):1207–1213
Menezes-Filho JA, Bouchard M, Sarcinelli PN, Moreira JC (2009b) Manganese exposure and the neuropsychological effect on children and adolescents: a review. Rev Panam Salud Publica 26 (6):541–548

Menezes-Filho JA, Novaes Cde O, Moreira JC, Sarcinelli PN, Mergler D (2011) Elevated manganese and cognitive performance in school-aged children and their mothers. Environ Res 111:156–163

Menezes-Filho JA, de Carvalho-Vivas CF, Viana GFS, Ferreira JRD, Nunes LS, Mergler D, Abreu N (2014) Elevated manganese exposure and school-aged children's behavior: a gender-stratified analysis. Neurotoxicology 45:293–300

Mergler D, Huel G, Bowler R, Iregren A, Bélanger S, Baldwin M, Tardif R, Smargiassi A, Martin L (1994) Nervous system dysfunction among workers with long-term exposure to manganese. Environ Res 64:151–180

Mergler D, Baldwin M, Belanger S, Larribe F, Beuter AR, Panisset M, Edwaeds R, De Geoffroy A, Sassine MP, Hudnell K (1999) Manganese neurotoxicity, a continuum of dysfunction: results from a community based study. Neurotoxicology 20:327–342

Molina-Villalba I, Lacasaña M, Rodríguez-Barranco M, Hernández A, Gonzalez-Alzaga B, Aguilar-Garduño C, Gil F (2015) Biomonitoring of arsenic, cadmium, lead, manganese and mercury in urine and hair of children living near mining and industrial areas. Chemosphere 124:83–91

Montes S, Riojas-Rodríguez H, Sabido-Pedraza E, Ríos C (2008) Biomarkers of manganese exposure in a population living close to a mine and mineral processing plant in Mexico. Environ Res 106(1):89–95

Mora AM, Arora M, Harley KG, Kogut K, Parra K, Hernández-Bonilla D, Gunier RB, Bradman A, Smith DR, Eskenazi B (2015) Prenatal and postnatal manganese teeth levels and neurodevelopment at 7, 9, and 10.5 years in the CHAMACOS cohort. Environ Int 84:39–54

Ntihabose R, Surette C, Foucher D et al (2018) Assessment of saliva, hair and toenails as biomarkers of low level exposure to manganese from drinking water in children. Neurotoxicology 64:126–133

Otero-Pregigueiro D, Fernández-Olmo I (2018) Use of CALPUFF to predict airborne Mn levels at schools in an urban area impacted by a nearby manganese alloy plant. Environ Int 119:455–465

Ranzi A, Fustinoni S, Erspamer L, Campo L, Gatti MG, Bechtold P, Bonassi S, Trenti T, Goldoni CA, Bertazzi PA, Lauriola P (2013) Biomonitoring of the general population living near a modern solid waste incinerator: a pilot study in Modena, Italy. Environ Int 61:88–97

Reis AP, Costa S, Santos I, Patinha C, Noack Y, Wragg J, Cave M, Sousa AJ (2015) Investigating relationships between biomarkers of exposure and environmental copper and manganese levels in house dusts from a Portuguese industrial city. Environ Geochem Health 37(4):725–744

Riojas-Rodríguez H, Solís-Vivanco R, Schilmann A, Montes S, Rodríguez S, Ríos C, Rodríguez-Agudelo Y (2010) Intellectual function in Mexican children living in a mining area and environmentally exposed to manganese. Environ Health Perspect 118(10):1465–1470

Rodrigues JLG, Araújo CFS, dos Santos NR, Bandeira MJ, Anjos ALS, Carvalho CF, Lima CS, Abreu JNS, Mergler D, Menezes-Filho JA (2018a) Airborne manganese exposure and neurobehavior in school-aged children living near a ferro-manganese alloy plant. Environ Res 167:66–77

Rodrigues JLG, Bandeira MJ, Araújo CFS, dos Santos NR, Anjos ALS, Koin NL, Pereira LC, Oliveira SSP, Mergler D, Menezes-Filho JA (2018b) Manganese and lead levels in settled dust in elementary schools are correlated with biomarkers of exposure in school-aged children. Environ Pollut 236:1004–1013

Rodríguez-Agudelo Y, Riojas-Rodríguez H, Ríos C, Rosas I, Sabido Pedraza E, Miranda J, Siebe C, Texcalac JL, Santos-Burgoa C (2006) Motor alterations associated with exposure to manganese in the environment in Mexico. Sci Total Environ 368(2–3):542–556

Rodríguez-Barranco M, Lacasaña M, Aguilar-Garduño C, Alguacil J, Gil F, González-Alzaga B, Rojas-García A (2013) Association of arsenic, cadmium and manganese exposure with neurodevelopment and behavioural disorders in children: a systematic review and meta-analysis. Sci Total Environ 454–455:562–577

Roels H, Lauwerys R, Buchet J, Genet P, Sarhan MJ, Hanotiau I, de Fays M, Bernard A, Stanescu D (1987) Epidemiological survey among workers exposed to manganese: effects on lung, central nervous system, and some biological indices. Am J Ind Med 11:307–327

Roels H, Ghyselen P, Buchet JP, Ceulemans E, Lauwerys RR (1992) Assessment of the permissible exposure level to manganese in workers exposed to manganese dioxide dust. Br J Ind Med 49:25–34

Roels H, Bowler RM, Kim Y, Henn BC, Mergler D, Hoet P, Gocheva VV, Bellinger DC, Wright RO, Harris MG, Chang Y, Bouchard MF, Riojas-Rodriguez H, Menezes-Filho JA, Téllez-Rojo MM (2012) Manganese exposure and cognitive deficits: a growing concern for manganese neurotoxicity. Neurotoxicology 33:872–880

Roth JA (2006) Homeostatic and toxic mechanisms regulating manganese uptake, retention, and elimination. Biol Res 39:45–57

Rugless F, Bhattacharya A, Succop P, Dietrich KN, Cox C, Alden J, Kuhnell P, Barnas M, Wright R, Parsons PJ, Praamsma ML, Palmer CD, Beidler C, Wittberg R, Haynes EN (2014) Childhood exposure to manganese and postural instability in children living near a ferromanganese refinery in Southeastern Ohio. Neurotoxicol Teratol 41:71–79

Santamaria AB (2008) Manganese exposure, essentiality & toxicity. Indian J Med Res 128:484–500

Santos-Burgoa C, Rios C, Mercado LA, Arechiga-Serrano R, Cano-Valle F, Eden-Wynter RA, Texcalac-Sangrador JL, Villa-Barragan JP, Rodriguez-Agudelo Y, Montes S (2001) Exposure to manganese: health effects on the general population, a pilot study in Central Mexico. Environ Res 85(2):90–104

Schwartz R, Apgar BJ, Wien EM (1986) Apparent absorption and retention of Ca, Cu, Mg, Mn, and Zn from a diet containing bran. Am J Clin Nutr 43(3):444–455

Smith D, Gwiazda R, Bowler R, Roels H, Park R, Taicher C, Lucchini R (2007) Biomarkers of Mn exposure in humans. Am J Ind Med 50:801–811

Solís-Vivanco R, Rodríguez-Agudelo Y, Riojas-Rodríguez H, Ríos C, Rosas I, Montes S (2009) Cognitive impairment in an adult Mexican population non-occupationally exposed to manganese. Environ Toxicol Pharmacol 28(2):172–178

Standridge JS, Bhattacharya A, Succop P, Cox C, Haynes EN (2008) Effect of chronic low level manganese exposure on postural balance: a pilot study of residents in Southwest Ohio. J Occup Environ Med 50(12):1421–1429

Strauss E, Sherman EMS, Spreen O (2006) A compendium of neuropsychological tests: administration, norms, and commentary, 3rd edn. Oxford University Press, New York

Torrente M, Colomina MT, Domingo JL (2005) Metal concentrations in hair and cognitive assessment in an adolescent population. Biol Trace Elem Res 104:215–221

Torres-Agustín R, Rodríguez-Agudelo Y, Schilmann A, Solís-Vivanco R, Montes S, Riojas-Rodríguez H, Cortez-Lugo M, Ríos C (2013) Effect of environmental manganese exposure on verbal learning and memory in Mexican children. Environ Res 121:39–44

US Department of Health and Human Services (2018) Fourth National Report on human exposure to environmental chemicals, vol 1. Updated tables, March 2018

US EPA (1984) Locating and estimating sources of manganese. https://www3.epa.gov/ttnchie1/le/manganes.pdf

US EPA (1993) Printout of reference concentration (RfC) for chronic inhalation exposure for manganese as verified 9/23/93, dated 12/93. Washington

US EPA (2012) Inhalation health effect reference values for manganese and compounds. EPA/600/R-12/047F5, December 2012

Viana GFS, Carvalho CF, Nunes LS, Rodrigues JLG, Ribeiro NS, Almeida DA, Ferreira JRD, Abreu N, Menezes-Filho JA (2014) Noninvasive biomarkers of manganese exposure and neuropsychological effects in environmentally exposed adults in Brazil. Toxicol Lett 231:169–178

Vimercati L, Baldassarre A, Gatti MF, Gagliardi T, Serinelli M, De Maria L, Caputi A, Dirodi A, Galise I, Cuccaro F, Assennato G (2016) Non-occupational exposure to heavy metals of the residents of an industrial area and biomonitoring. Environ Monit Assess 188:673

Winder BS, Salmon AG, Marty MA (2010) Inhalation of an essential metal: development of reference exposure levels for manganese. Regul Toxicol Pharmacol 57:195–199

World Health Organization (2005) Manganese and its compounds: environmental study

Zheng W, Fu SX, Dydak U, Cowan DM (2011) Biomarkers of manganese intoxication. Neurotoxicology 32:1–8

Zoni S, Albini E, Lucchini R (2007) Neuropsychological testing for the assessment of manganese neurotoxicity: a review and a proposal. Am J Ind Med 50:812–830

Photodegradation of Pharmaceutical and Personal Care Products (PPCPs) and Antibacterial Activity in Water by Transition Metals

Godfred Kwesi Teye, Williams Kweku Darkwah, Huang Jingyu, Li Ke, and Yi Li

Contents

Abstract The intensity of emerging pollutants such as pharmaceuticals and personal care products (PPCPs) in the aquatic and terrestrial environment is a major source of concern to researchers. The current conventional methods of wastewater treatment plants are considered not efficient enough in the complete removal of the

---

Huang Jingyu and Yi Li equally contributed to this work.

G. K. Teye · W. K. Darkwah · H. Jingyu (✉) · Y. Li
Key Laboratory of Integrated Regulation and Resource Development on Shallow Lakes, Ministry of Education, College of Environment, Hohai University, Nanjing, People's Republic of China
e-mail: teyegodfred@hhu.edu.cn; williams.darkwah@stu.ucc.edu.gh; huangjingyu120120@126.com; envly@hhu.edu.cn

L. Ke
Department of Civil Engineering, Jilin Jianzhu University, Jilin, People's Republic of China
e-mail: like_student@163.com

P. de Voogt (ed.), *Reviews of Environmental Contamination and Toxicology Volume 254*, Reviews of Environmental Contamination and Toxicology 254, https://doi.org/10.1007/398_2020_47

recalcitrant contaminants from water. The use of modified transition metals in visible responsive synthesis to degrade PPCPs and other pollutants (organic and inorganic) is considered as a developing green chemistry and sustainable technology. Hence, this review presents the state-of-the-art discussion on the novel photodegradation of PPCPs, and antibacterial activities of transition metal-modified magnetite materials for wastewater treatment, and suggested directions for the future. Transition metal-modified magnetite nanostructured photocatalysis is identified as one of the best candidates employed in advanced oxidation processes (AOPs) for wastewater treatment and has been found to efficiently destroy bacterial spores and effectively remove recalcitrant pollutants in water. Therefore, this article hopes to contribute scientific knowledge along with existing ones on advanced mechanisms and technology used in wastewater treatment.

Keywords Advanced oxidation processes · Photocatalytic · PPCPs · Transition metals · Wastewater treatment

Abbreviations

AOP	Advanced oxide processes
$BIPO_4$	Bismuth phosphate
C3N4	Carbon nitride
Cu3P	Copper phosphide
DCF	Diclofenac
EDCs	Endocrine disruption chemicals
FeO	Iron(II) oxide
IUPAC	International union of pure and applied chemistry
NM	Noble metal
NSAIDS	Nonsteroidal anti-inflammatory drugs
OTC	Over-the-counter
PPCPs	Pharmaceutical and personal care products
SA	Salicylic acid
TBHP	Tert-butyl hydroperoxide
TiO	Titanium oxides
UV	Ultraviolet light
WWTPs	Wastewater treatment plants
ZnO	Zinc oxides

1 Introduction

Wastewater treatment and reuse are essential approaches to ease pollution and water shortage (Tang et al. 2018). For safe water reuse applications, effective removal of a wide range of pollutants including bulk organics, salts, nutrients, and contaminants is essential. Among these pollutants, the effective removal of contaminants such as pharmaceuticals, pesticides, steroid hormones, industrial chemicals, and ingredients of personal care products is one of the most challenging aspects of wastewater treatment (Hai et al. 2014; Radjenović et al. 2009; Asif et al. 2018). Since contaminants are potentially harmful to the aquatic ecosystems and human health (Alexander et al. 2012; Hai et al. 2018; Schwarzenbach et al. 2006), the development of a wastewater treatment process for effective removal of contaminants has gained significant interest in the recent years. Photocatalytic degradation is considered a zero-waste procedure to treat wastewater and environmental pollutants in the use of nano-based catalysts such as TiO_2 and ZnO, among others (Durairaj et al. 2018).

There are progressive and rigorous environmental standards as well as evolving contaminants such as persistent organic pollutants of pharmaceutical origin and hormones which have forced research communities to look for new treatment technologies to ensure sustainable supply of potable water to society (Leong et al. 2014). This is because the pollution of water has become a crucial worldwide menace because it reduces quality, causes environmental degradation, and introduces risk to public health (Akkari et al. 2018). The use of transition metals in the photocatalytic process to degrade pollutants is considered as green chemistry and a sustainable technology (Zhou et al. 2018).

Zhu and Wang (2017) described photocatalysis as a branch of research which focuses on how to use the energy delivered by photons instead of heat to enable and ensure powerful chemical reactions (Zhu and Wang 2017). This technology is a very significant intervention that contributes to solving pertinent issues relating to environmental management, particularly pollutants in wastewater media. Over the years, several efforts have been made to develop sustainable technologies to help improve upon the treatment methods for contaminants in wastewater (Borges et al. 2016; Fung et al. 2019; Yang et al. 2012). Borges and his colleagues in their studies stated that the heterogeneous photocatalysis belongs to the advanced oxidation processes (AOP). They further posited that the technology is versatile, economical, a promising green chemistry, environmentally friendly, and useful, based on pollutants present in gaseous and aqueous effluent (Borges et al. 2016). Also, some levels of success have been demonstrated in the ability of AOP to remove low levels of recalcitrant contaminants as well as microorganisms (Borges et al. 2016; Sharma et al. 2016; Miklos et al. 2018; Deng and Zhao 2015).

Other studies further considered that among the various treatment methods for the removal of pollutants such as PPCPs from effluent and receiving water including conventional activated sludge, soil aquifer treatment, AOP, and biomembrane process, the latter is one of the most effective processes (Wang et al. 2018a, b). Wastewater streams currently contain several ranges of contaminants as was found by Schymanski et al. (2014), Tran et al. (2018), and Reemtsma et al. (2016). They established the presence of a number of pollutants flowing unreservedly in the environment. This basically affects ecological safety and life on earth as a result of agricultural, industrial, and domestic activities (Khan et al. 2017). The pollutants need to be removed in order to protect and maintain life on earth and that of the aquatic ecology. The purification of the ecosystem, mainly contaminated water media, is highly important for the survival of human life on earth (Fujishima et al. 2013). Some of these contaminants described here include sulfonamides and pharmaceutical and personal care products (PPCPs) (Sharma et al. 2016).

PPCPs are a major group of environmental pollutants (Savun-Hekimoğlu and Ince 2017), which are mostly present in aqueous effluent that needs to be eliminated or reduced to the lowest form in order not to cause harm to human life as well as other organisms (Boxall et al. 2012; Cizmas et al. 2015; Wang and Wang 2016a, b). For instance, commonly used PPCP such as salicylic acid (SA) is a bio-persistent compound and toxic to aquatic organisms, produced by hydrolytic deacetylation of acetylsalicylic acid and used as ingredient of acne treatment, moisturizers, and facial cleaners, among others (Savun-Hekimoğlu and Ince 2017). There is also growing concern that sulfonamides which are one of the groups of PPCPs have adverse health implications on human life if not thoroughly treated (Wang et al. 2018a, b). PPCPs are a class of evolving contaminants regularly used as medicinal, cosmetic, and personal care products (Wang et al. 2018a, b). Bu et al. (2013) and Richardson et al. (2005) revealed in their study that there's mounting evidence that these contaminants have grave consequences on human health as well as lower living organisms. Other studies also pointed to the fact that PPCPs and sulfonamides in aquatic medium are constantly being discharged and they often contain traces of contaminants which have the tendency to bioaccumulate in aquatic organisms with a potential threat to human life (Ebele et al. 2017; Muir et al. 2017; Chen et al. 2017).

Research has been carried on many photocatalysts such as TiO_2, C_3N_4, and $BIPO_4$, among others, because it is believed to be economically good, environmentally friendly, and safe in water (Zeng et al. 2019; Yi et al. 2019; Djouadi et al. 2018). Comparatively, not much has been done on ZnO because of its polycrystalline nature (Shon et al. 2008; Adormaa et al. 2018; Shi et al. 2015). Gaya et al. (2009) indicated that TiO_2, the chief semiconductor, used due to its inertness both chemically and biologically, is efficient and economically viable but has some shortcomings to the application in wastewater remediation when used on a large scale. They therefore suggested that there should be a renewed interest in the search for more reliable photocatalysts. They concluded that ZnO demonstrates relatively equivalent attributes and flexibility to TiO_2 (Gaya et al. 2009). The quest for renewed interest to add to knowledge on the photocatalytic activity of ZnO is one of the motivating factors for this review.

One of the objectives of this review is to critically review current knowledge on photocatalytic activity of transition metals in relation to PPCPs in wastewater treatment and to suggest directions for future research. The review, therefore, gives an overview of the types of transition metals and PPCPs. It further focuses on broad definitions of photocatalysis in terms of synthesis. The authors discussed recent studies on the visible light harvest activities of modified transition metals in relation to PPCPs photodegradation performances, as well as antibacterial activity and photocatalytic principles of transition metals. The review further outlines the modern studies on photodegradation of PPCPS using transition metal-modified magnetite nano-materials and related pollutants and concluded with a look at the application of transition metals, photocatalysis, and the future perspective.

2 Transition Metals

Photocatalytic activities of most photocatalysts in visible light synthesis are usually enhanced by cocatalysts such as transitional metals by promoting the transfer and separation efficiency of electron-hole pairs. A work led by Chen (Qin et al. 2018) indicated that the transitional metals such as Cu_3P, ZnO, FeO, and TiO coupled with nano-based particles, for instance, n-type graphitic carbon nitride, usually forms a p-n junction to promote charge separation and transfer for an improved photocatalytic activity.

According to Habashi and City (2017), “transition metals are the metals which are scientifically placed in the upright groups on the periodic table from scandium to nickel.” These metals have electronic configuration which is different from the inert gases. They are, however, categorized based on the fact that they have the same number of electrons in their outmost shells and a steadily increasing number of electrons in the inner shell. This is as a result of energy levels which are determined from spectroscopic measurements. This deals with the intensity of radiation as a function of wavelength. However, the International Union of Pure and Applied Chemistry (IUPAC) defines “transition metals as elements with an incomplete d subshell or elements that can form a cation with an incomplete d subshell” (Strohfeldt 2015).

They, therefore, have the characteristics and properties as nearly as all the d-block metals, for instance, very hard, ductile, and pliable and conduct electricity and heat. Oftentimes, they form one of the typical metal structures with the exceptions of manganese (Mn), zinc (Zn), cadmium (Cd), and mercury (Hg). For example, iron (Fe), cobalt (Co), and nickel (Ni) are the only three d-block elements known to produce a magnetic field. The above properties make them interesting to use in the photocatalytic processes. There are five (5) key properties the transition metals possess (Habashi and City 2017). These include but not limited to the following;

(a) Their appearance is quite close besides showing the usual group affiliations, because they have the same quantity of the remotest electrons.
(b) Sometimes they might lose additional electrons from the next lower shell to form ions with higher charges.
(c) The infinitesimal radius of the following metals in a particular horizontal period declines slightly as the atomic number increases because when an electron is added to an inner shell, it decreases slightly in size of the atom as a result of increased electrostatic attraction.
(d) Mostly, they form colored ions in a mixture because of electronic modifications and form covalent compounds, for instance, the carbonyls of iron and nickel.
(e) They finally form borides, carbides, nitrides, and hydrides, which have mostly metallic character, and also form coordination composites with ammonia, for instance, the amines of cobalt and nickel.

Transition metals are very important in the management of the environment especially those which are used in the photocatalytic processes. This was confirmed by Deng et al. (2018) in their research which indicated that noble metal

(NM) catalysts have significant application prospects in the chemical production processes and environmental protection fields. For instance, to ensure the reduction of harmful metals to the environment as well as human health, the levels of excessive aromatics needed to be reduced through catalytic hydrogenation (Deng et al. 2018).

3 Pharmaceutical and Personal Care Products (PPCPs)

PPCPs in the aquatic environment, for instance, surface water, groundwater, as well as in wastewater effluents, continue to constitute an evolving issue of concern due to the lack of adequate information in relation to their occurrence, fate, and ecotoxicological effects in sewage, surface, ground, and drinking water. Studies have shown that the use of PPCPs extending from pain relievers, antibacterial soap, to over-the-counter (OTC) drugs have increased in recent times, with large quantities competing closely with that of agrochemicals, (Kagle et al. 2009; Savun-Hekimoğlu and Ince 2017; Esplugas et al. 2007). The increase in the use of PPCPs and their subsequent presence in the environment requires understanding of how these compounds are bio-transformed and their derivative effects on humans and other biotic factors. The global usage of PPCPs has surged up in the past decades. Tijani et al. (2016) found that developed and emerging economies such as Russia, Brazil, India, South Africa, and China are among the leading consumers.

According to Kagle et al. (2009), PPCPs are a group of compounds which collectively refer to any substance used for medicinal or personal care purposes. They may include metabolites and transformation products of these substances. For example, a number of PPCPs and their metabolites (metabolites of nonionic surfactants) have been shown to have toxicological effects and are estrogenic. Tijani et al. (2016) and Awfa et al. (2018), on the other hand, described PPCPs as synthetic persistent substances that are envisioned for human and animal healthcare and medical purposes. They indicated that PPCPs can further be classified as antibiotics, anticonvulsants, contrast agents, hormones, nonsteroidal anti-inflammatory drugs (NSAIDs), b-blockers, lipid regulators, painkillers, preservatives, disinfectants, insect repellants, fungicides, soaps and detergents, fragrances, and sunscreen UV filters. Whatever the situation, remedial action is required to clean the environment in order to break the chain of the toxicological effects (Luo et al. 2014). Understanding of photocatalysis biodegradation could aid in reduction of the concentration of PPCPs in the environment, both in aquatic and terrestrial medium. The PPCPs have been found to enter the environment through agricultural and crop irrigation in the form of micro-pollutants as a result of the use of reclaimed wastewater. Conventional wastewater treatment processes are moderately ineffective as far as the removal of PPCPs and wastewater-related organic pollutants is concerned. This is because most of the compounds are intrinsically biotic in nature (Miller et al. 2015). Jiang et al. (2013) and Lester et al. (2013) concluded that the presence of PPCPs in surface waters is largely due to incomplete removal of PPCP pollutants by sewage treatment plants and industrial wastewater treatment plants.

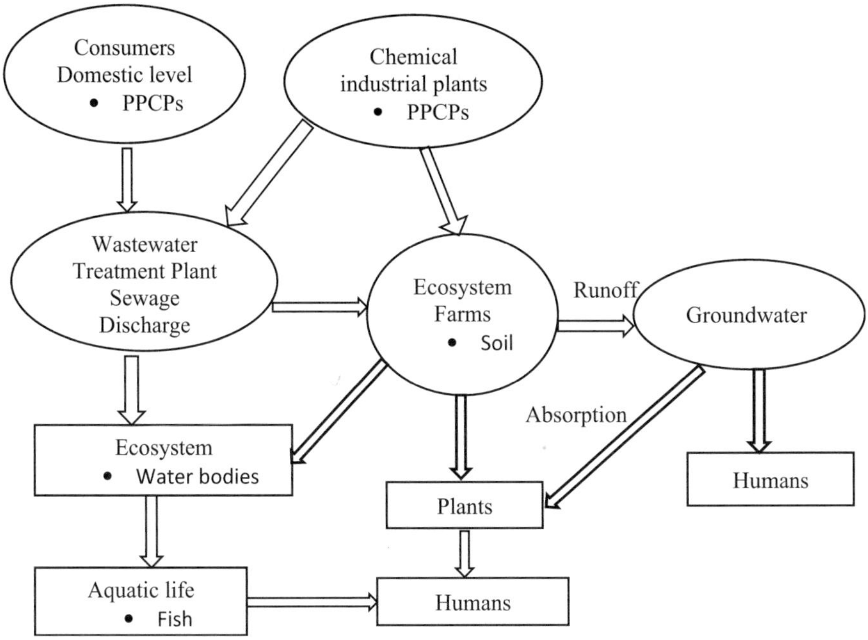

Fig. 1 Routes of entry of PPCPs into the environment

Freyria (2018) in their review on nano-materials for the abatement of PPCPs from wastewater posited that concentrations of PPCPs are continuously introduced into several ecosystems ranging from soil, groundwater, to freshwater after consumption. This happens through sewage and wastewater treatment plants (WWTPs), which are finally found in watercourses and in their surrounding environment (Freyria 2018). This gives an indication that Miller et al. and Freyria agree on the routes of entry of PPCPs into the environment. When PPCPs are discharged from wastewater treatment plants and agricultural runoff, their routes of entry becomes constant, and these pollutants are very tiny to be easily detected, analyzed, and degraded during the treatment of water (Luo et al. 2014)

Figure 1 illustrates the routes of entry into the environment by PPCPs as discharged from domestic or household consumers into the wastewater treatment plants as well as improperly managed industrial sewerage systems.

The PPCPs permeate the ecosystem as runoff from farms through animal care and crop irrigation. The wastewater treatment plants are inefficient in the removal of PPCPs (Fernández et al. 2014), so the pollutants enter the ecosystem through freshwater bodies, soil, underground water bodies, and sediment runoff where they are absorbed by roots of plants as well as consumed by aquatic lives. This was established by Kagle et al. (2009) that most PPCPs enter the wastewater treatment plants (WWTPs) as a constituent of domestic, hospital, and industrial waste or through runoff. This is because huge quantities of medication taken by patients pass through their body system unchanged and move through urine and fecal matter into the wastewater. The sludge produced in WWTPs is also normally applied in everyday farmlands as compost for soil improvement. The presence of pharmaceuticals in soils with low sorption affinity can contribute to the contamination of adjacent surface and groundwaters through runoff and permeation. Kagle et al. (2009) suggested that the number of PPCPs being detected in WWTPs is on the the rise due to the growth of pharmaceutical and personal care industries as well as the improvement on the systems of detection. Humans in turn consumed aquatic organisms and the plants, and through the process of bioaccumulation and biomagnification, they are variously transferred into the body parts of humans causing resistance in relation to antibiotics such as sulfonamides as indicated by Khraisheh et al. (2014).

These pollutants seem to have a wide spread consumer base since they could be found in most PPCPs. In support of the above, Li et al. (2017) in their previous investigations affirmed that effluents from wastewater treatment plants (WWTPs) are significant sources of these emerging contaminants in the environment. The design of conventional WWTPs is normally intended to remove organic matter, nitrogen, and phosphate, but not the emerging compounds. This could probably account for the reason why most conventional plants are not able to remove PPCPs as was found by Khraisheh et al. (2014); some of these products range from antibiotics, toothpaste, deodorant, soaps, polymer, and fibers. The wide spread of PPCPs and in some instances endocrine disruption chemicals (EDCs) in the WWTP effluents as well as receiving aquatic media may affect the quality of water. It also poses a potential danger to public health as well as other living organisms in the aquatic environment. These effluents have been described as a prospective source of micro-pollutants for water environment (Yang et al. 2012).

The conventional pollutants relating to emerging pollutant intensities are usually low, ranging between ng/L and mg/L. But their toxicity and associated challenges, for example, antibiotic resistance, may cause potential risks to human health and the environment (Li et al. 2015). The continuous discharge of PPCPs into the ecosystem is a serious challenge. For instance, wastewater containing a mix of pharmaceuticals caused significant reproductive and other health impacts for fish, and other aquatic mammals, and also PPCPs, such as acetaminophen, caffeine, 1,7-dimethylxanthine, dehydronifedine, tetracycline, oxytetracycline, sulfonamides, macrolides, and ormetoprim, have been detected in fruits, vegetables, meat, and milk due to uptake from a PPCP-polluted environment (Awfa et al. 2018). The public health repercussions of these pollutants on human life and that of aquatic ecology could be phenomenal, and therefore, there is the need to adopt an efficient measure to improve upon the available wastewater treatment systems.

4 Global Production of Drinking Water

In 2004, it was estimated that about 15% of the world's population, mostly living in the less- favored regions of the planet, did not have access to enough fresh water to satisfy their daily needs, and this number was expected to double by 2015 (Villén et al. 2006). This represents a serious public health issue since waterborne, water-washed, and water-based diseases are associated with lack of improvement in domestic water supply and adequate sanitation (Villén et al. 2006). Development of low cost-effective methods, such as photocatalysis, for removal of pollutants from water supplies can help alleviate this problem, especially in communities where water disinfection must have sufficiently low operational costs. Alternative technologies to traditional chlorination are now being considered for household use (Najm and Trussel 1999). Solar disinfection (SODIS) using nano-based particles is a simple technology that is capable of inactivating many waterborne pathogenic bacteria using the combined effect of solar UV-A radiation and temperature (Dellarocas et al. 2009). This method is of low cost and does not produce toxic byproducts; however, it limits the volume of water subjected to treatment (typically 2 L per exposed water bottle) and has a disadvantageous long time of process (typically 2 day exposure for complete inactivation) (Villén et al. 2006). The combination of sunlight and photocatalyst is a promising option for water treatment in areas with insufficient infrastructure but high annual sunshine. The use of compound parabolic reactors as an efficient technology to collect and focus diffuse and direct solar radiation onto a transparent pipe containing contaminated water has demonstrated feasibility to disinfect water using TiO_2 suspensions (Mcloughlin et al. 2004). The European Union International Cooperation program (INCO) has sponsored initiatives for developing a solar photocatalysis-based cost-effective technology for water decontamination and disinfection in rural areas of developing countries. These projects are the SOLWATER and AQUACAT projects, respectively. They are aimed at developing a solar reactor to decontaminate and disinfect small volumes of water. So far field tests with the final prototypes were carried out to validate operation under real conditions (Navntoft et al. 2007). The final SOLWATER prototype was composed of two tubes containing Ahlstrom paper impregnated with titanium dioxide and two tubes containing a supported photosensitizer (Villén et al. 2006). These tubes are placed on a compound parabolic concentration collector and run in series, where electricity is provided by a solar panel.

Field tests using the SOLWATER prototype involved placement of the reactor in the yard of a shanty house in Los Pereyra, Tucuman, Argentina. It studied the removal of bacterial contamination during 3 months of testing, using natural water contaminated with coliforms, *E. faecalis*, and *P. aeruginosa*, as well as high levels of natural organic matter and variable inorganic pollutants (Navntoft et al. 2007). The SOLWATER reactor was shown to be effective for this application. Similar tests have been performed in photoreactors installed in various geographic regions, including Egypt, France, Greece, Mexico, Morocco, Peru, Spain, Switzerland, and Tunisia (Villén et al. 2006). Other research in the field of potable water production in developing countries include the development of affordable and efficient technology. It should be noted that there has also been significant research done in the advance of solar disinfection and photodegradation of these highly resistant pollutants such as PPCPs using SODIS alone.

5 Photocatalysis

In recent years, the use of advanced oxidation processes (AOPs) in the treatment and disinfection of wastewater has grown. Many countries including Canada, Germany, Ireland, and other parts of Europe are now employing tertiary treatment steps to provide a higher quality of water. The most popular AOP employed is UV, which has been found to efficiently disinfect water and can effectively remove micropollutants such as PPCPs depending on the lamp used (Liang et al. 2019). The application of UV is one of the easiest treatments to integrate as it can be retrofitted into existing chlorine contact chambers with the use of certain lamps. However, its application results in a slightly higher cost, but it is more effective in disinfection than the chlorination and dechlorination processes (O'Connor et al. 2012).

At present, the application of photocatalytic AOPs (heterogeneous and homogeneous) in water treatment is still a work in progress despite hundreds of literature articles published every year documenting its efficiency for the elimination of various organic compounds (Zhao et al. 2018; Czech and Tyszczuk-Rotko 2018; Chen et al. 2011; Ali et al. 2018; Darkwah et al. 2020). There are also numerous research centers worldwide which have large-scale solar photocatalytic equipment, e.g., the Plataforma Solar de Almería (PSA) in Spain. However, it may take some time before systems like these are introduced into actual water treatment facilities worldwide (Wang and Wang 2016a, b).

During heterogeneous photocatalysis process, the catalyst is activated not by thermal methods but by photonic methods (Fig. 2). Hermannsdörfer et al. (2012) indicated that at present, there is no current definition by the IUPAC for heterogeneous photocatalysis although it has been aptly described by authors as "photo-induced molecular transformations or reactions which take place at the surface of a catalyst"(Linsebigler et al. 1995).

6 Mechanism of Photocatalysis

The fundamental processes of heterogeneous photocatalysis consist of five stages. The first stage comprises of the transfer of the reactants to the surface of the semiconductor photocatalyst (Figs. 3 and 4).

In the second stage, adsorption of at least one of the reactants or the sorbate transpires. The third stage in classical heterogeneous catalysis is a simple reaction of

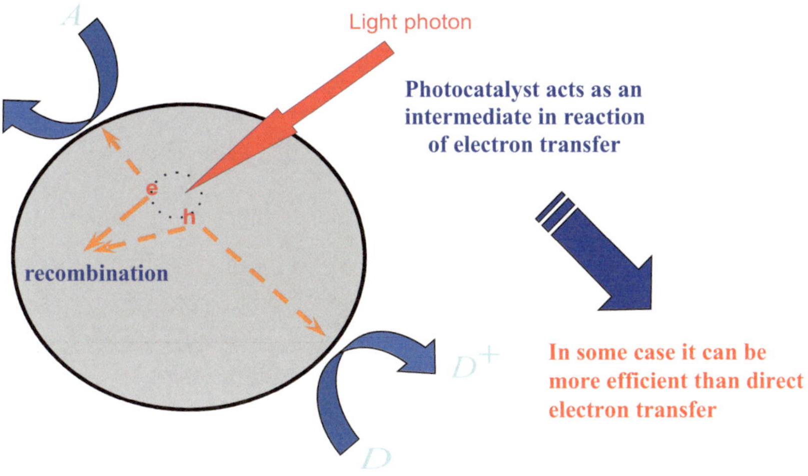

Fig. 2 Scheme of Redox reactions driven by light in photocatalyst. Reproduced with permission from Darkwah et al. (2020). Copyright 2020, Royal Society of Chemistry

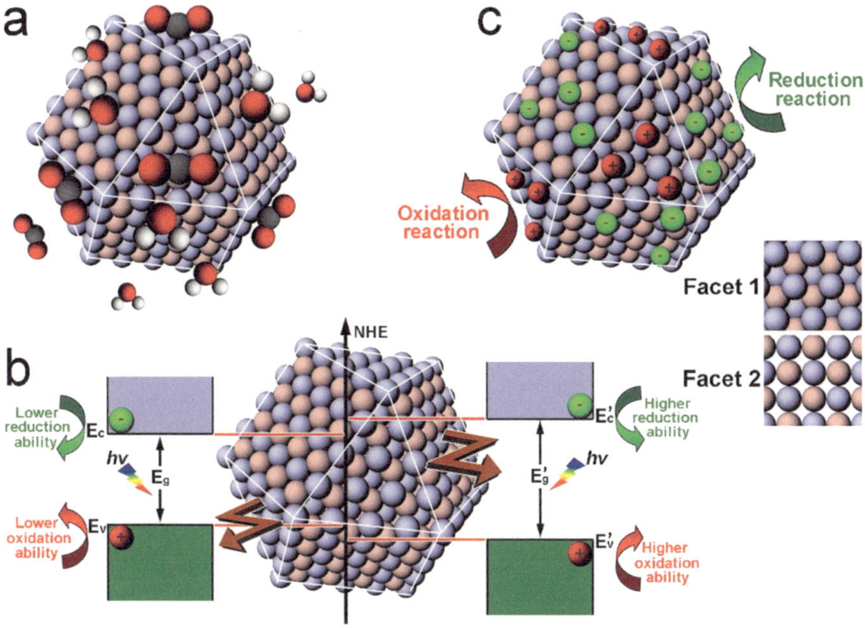

Fig. 3 Important roles of facets in the surface design of photocatalytic materials. (**a**) Adsorption and activation of reactant molecules on different facets. (**b**) Redox abilities of photogenerated charge carriers tuned by the surface electronic band structures of different facets. (**c**) Accumulations of photogenerated electrons or holes on different facets. Adapted with permission from Bai et al. (2017). Copyright 2017, Wiley-VCH Verlag GmbH & Co. KGaA, Weinheim

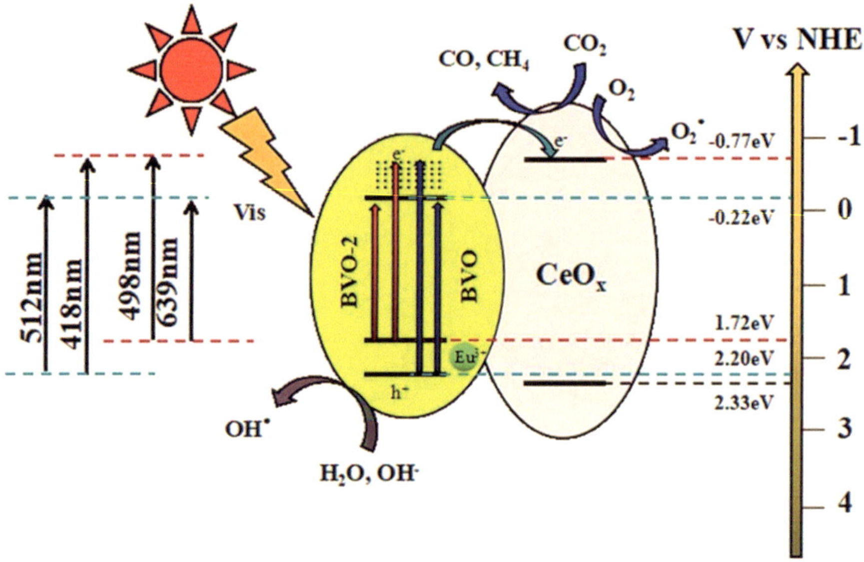

Fig. 4 Schematic of transfer and separation of photogenerated high-energy electrons of 2C/BVO-2 nanocomposites under visible illumination. Reproduced with permission from Xu et al. (2019b). Copyright 2019, Royal Society Chemistry

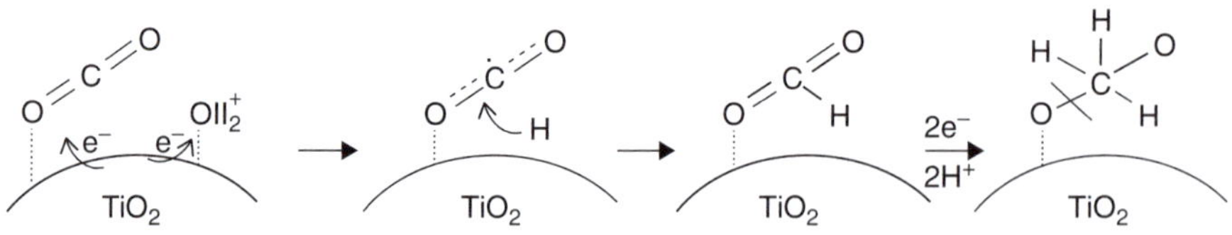

Fig. 5 Mechanism of photoreduction of CO_2 to methoxyl radical on TiO_2 in the presence of water during photocatalytic reaction. Reproduced with permission from (Adormaa et al. 2018) Copyright 2018, Royal Society Chemistry

the adsorbed products; however, in heterogeneous photocatalysis, a stage which is much more complex involves three main mechanisms;

1. The simultaneous absorption of photons by the semiconductor and the reaction in adsorbed state
2. The creation of electron-hole pairs
3. The various redox processes including radical propagation, charge neutralization, and radical recombination

The fourth stage, on the other hand, contains desorption of the products (Figs. 4 and 5). The final stage encompasses the removal of the products from the interface region (Hermannsdörfer et al. 2012).

7 Treatment of PPCPs in Wastewater by Transition Metal-Modified Magnetite Photocatalysis

The ability of photocatalysis to break down and detoxify harmful organic chemicals such as PPCPs has been exploited for groundwater treatment (Table 1), as shown by engineering-scale demonstrations using solar photocatalysis. The disinfection abilities of photocatalytic processes (Figs. 5 and 10) for application to treating groundwater contaminated with microorganisms such as *F. solani* have been investigated and confirmed to be effective for the removal of such microorganisms (Fernández-Ibáñeza et al. 2009). Natural well water containing the *F. solani* species and solar illumination and employing solar CPCs was also explored as a process configuration for this application in treatment of PPCPs.

Transition metal photocatalysis has been proven to be effective in the removal of chemical compounds such as PPCPs and microbiological pathogens from water (Villén et al. 2006). A thorough review of the application of photocatalysis for the removal of biological species in this context by McCullagh et al. (2000). McCullagh et al. (2000) examines the disinfection and photodegradation of PPCPs outcome of transition metal suspensions, effect of additives and pH, respectively, on the photocatalytic abilities and disinfection effect of transition metal thin films, and the effect of electrochemically applied potential on the photo bactericidal effect of transition metal thin films.

The properties of catalysts such as crystallite dimensions, surface area, and the presence of hydroxy surface groups are revealed to be decisive to the photocatalytic proficiency of the transition metal-modified nanostructured materials (Silva et al. 2013). Silva et al. (2013) also studied the photochemical and photocatalytic degradation of the emerging pollutant trans-resveratrol (Fig. 6) under different irradiation wavelengths and using different TiO_2 catalysts and concluded that trans-resveratrol was easily degraded when irradiated using the whole spectral range (UV-Vis) rather than with UV and near-UV to visible irradiation. They further indicated that the photocatalytic process was more efficient than the pure photochemical one leading to higher degradation rates and higher organic content removal (Silva et al. 2013), as illustrated in Fig. 6.

Recent investigation by Zhang et al. (2019) on double metal-free polymer reactive sites for the efficient degradation of diclofenac by visible light-driven oxygen reduction to superoxide radical and hydrogen peroxide revealed that photocatalytic degradation of pharmaceutical products such as diclofenac under visible light demonstrated an excellent performance (Fig. 7).

Also, Barrocas et al. (2018) studied the photocatalytic degradation ability of modified titanate nano-based material on pharmaceutical products in wastewater. Their work focused on synthesis of new hybrid photocatalytic nanomaterials through sensitization of titanate nanotubes and nanowires with ethylenediamine. After 60 min under UV-vis radiation, the titanate nanotube (NTNT) sample was the best catalyst achieving 98% of photodegradation efficacy for phenol and theophylline and 60% for caffeine. These results cannot be attributed only to surface area

Table 1 Typical transition metals in photocatalysis

Metals	Properties	Modified metals	Modification schemes	Method of synthesis	Photocatalytic activity	Reference
Ti		TiO_2	$g\text{-}C_3N_4\text{-}N\text{-}TiO_2$ heterojunction	Facile in situ synthesis, thermal treatment	Excellent photoreduction of CO_2 to CO	Zhou et al. (2016)
Fe		Co-Fe nitrides	Coupling	Thermal treatment	Enhanced CO_2 reduction	Niu et al. (2015a, b)
Au		Modified Au (111) by assembling 4-mercaptopyridine (4MPy) and 4-aminothiophenol (4ATP)		Chemically functionalized with substituted and unsubstituted Fe(II)-phthalocyanines technique	Excellent electrocatalytic activity for the oxidation of L-cysteine	Teixeira et al. (2018); Gonzalez-Zavala et al. (2016); Xu et al. (2019a)
Mn		$Mn_3O_4\text{-}MnO_2$ hetero-nanorods/graphene nanocomposite (MM-HNRS/GS)	Heterojunction	Prepared by merging a hydrothermal method, a wet chemical process, and a thermal treatment, and X-ray diffraction is adopted to demonstrate the formation process	Outstanding synergistic influence and evidently improved the electrocatalytic activity	Darkwah et al. (2019)
Sn		$SnO_{2-X}/g\text{-}C_3N_4$	Heterojunction		An enhanced CO_2 reduction	Darkwah et al. (2019)
Zn		$g\text{-}C_3N_4/ZnO$	Heterojunction		An enhanced CO_2 reduction	Yu et al. (2017)
Ti	g-C3N4 and TiO2, under the same experimental conditions are surface of diatomite	TiO_2			Excellent degradation of (A) rhodamine B (RhB) and methylene blue (MB) and (B) methyl orange (MO) and phenol under the same conditions	Chen et al. (2019); Mathis et al. (2016); Xu et al. (2008); Zhao et al. (2019)
Fe		Fe_2O_4/ZnO	Heterojunction	Thermal method	Showed and enhanced degradative activity	Benaboud et al. (2017)
Zr		ZrO2		Soft chemical synthesis		Niu et al. (2015b)
Si		SiO2		Soft chemical synthesis		Li et al. (2015)

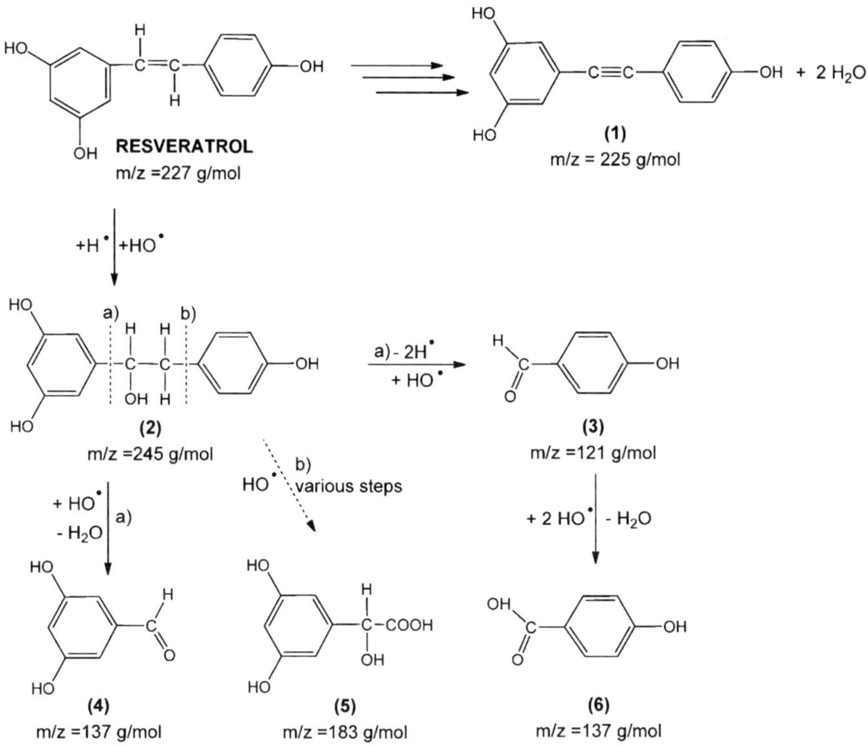

Fig. 6 Mechanism for direct and photocatalyzed degradation of trans-resveratrol. Adapted with permission from Silva et al. (2013). Copyright 2013, Elsevier

differences since NTNT presents the lowest surface area but allows to conclude that the amine sensitization leads to an enhancement of the sample's catalytic performance for this reaction. This suggests an active role of the surface amine groups in this process and permits anticipation of best results for amine-modified samples especially in photocatalytic applications where $OH^{\cdot}$ plays a crucial role.

These authors further proposed the mechanism for charge-transfer in irradiated NTNT for PPCPs photodegradation. During the charge-transfer processes in the TNT/TNW and NTNT/NTNW nanoparticles, the irradiated TNT/TNW and NTNT/NTNW nanoparticles, when activated, generate electrons (e^-) and holes (h^+) in the conduction (CB) and valence band (VB) of the semiconductor (see Fig. 5 for illustration). This process happens because these semiconductors have energies higher than their bandgap when activated.

The photogenerated holes will react with OH^- and/or HO oxidizing them to $OH^{\bullet}$ radicals. Simultaneously, the formation of $^{\bullet}O2$ species arises from the adsorbed oxygen reduction (Yasmeen et al. 2019a; Zada et al. 2019). The extension of this process will combine action of the photogenerated h^+, and others highly determine the decrease on the recombination rate. The oxidant species, including O2 and OH

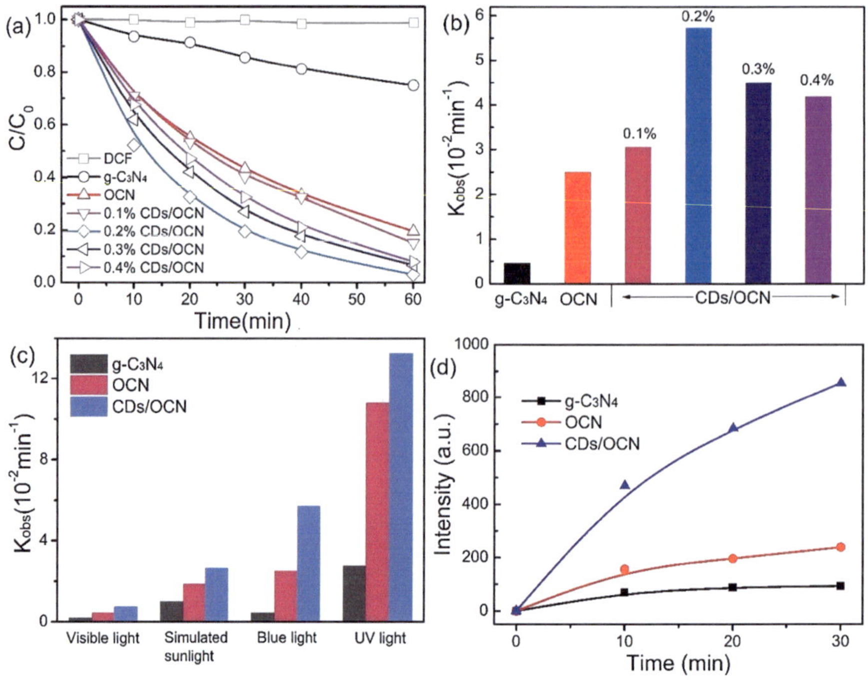

Fig. 7 Photocatalytic degradation of diclofenac (DCF) by the photocatalysts. (**a**) Under visible light irradiation, (**b**) kinetic rate constant of DCF over the photocatalysts, (**c**) different light source irradiation, and (**d**) photocatalytic generation of H_2O_2 in pure water under visible light irradiation with air atmosphere. Data obtained with permission from Zhang et al. (2019). Copyright 2019, Royal Society of Chemistry

radicals, will be responsible for the pollutants' photodegradation. If no O2 is present during this process, no $^{\bullet}$O2 species will be formed, and the photogenerated charge carriers' recombination will be higher. Therefore, much less holes will be available to oxidize OH^-/H2O for $OH^{\bullet}$ production (Yasmeen et al. 2019a). Consequently, no photocatalytic effect will be noticed, and the pollutant degradation will be only due to photolysis effect (Yasmeen et al. 2019a, b).

Studies of photocatalytic performance have clearly revealed that the nanoporous TiO_2 films (modified transition metal) unveiled an excellent UV photocatalytic activity for the degradation of PPCPs (Xu et al. 2015).

To compensate the cost there is the need for further enhancement of the photocatalyst. Two approaches based on photocatalytic principles have been developed to include the following:

1. Promoting the separation and transfer of photogenerated charge carriers (Fig. 8).
2. Extending the absorption wavelength to the visible light region. At present, the modifications that enhance the visible light photocatalytic activity of transition metals may be more attractive as they can utilize a large portion of sunlight. But in

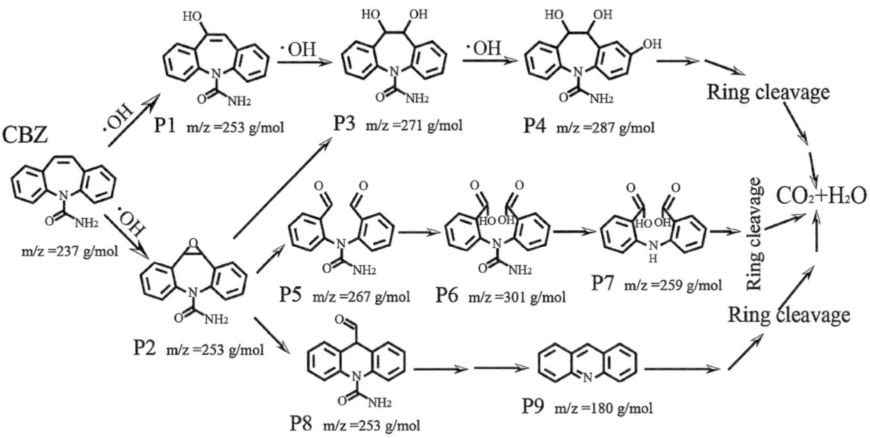

Fig. 8 Proposed degradation pathway of CBZ in the presence of $BiPO_4$. Reproduced with permission from (Guo et al. 2018; Darkwah et al. 2019). Copyright 2018, Elsevier & Royal Society of Chemistry

particular, these modifications should not decrease the UV photocatalytic activity at the same time (Darkwah et al. 2019); otherwise the activity under the real sunlight irradiation (UV + Vis) cannot be enhanced. The activity of these metals itself could be enhanced by improving the methods of synthesis through optimizing the structural parameters. These effective modification methods used to modify TiO_2 include doping, formation of the heterojunction structure, surface hybridization, and creation of the surface oxygen vacancies, among others (Darkwah et al. 2019).

Guo et al. (2018) and Wang et al. (2018a, b) studied the photocatalytic activity of twinned $BiPO_4$ by coupling with BiOCl. They reported that the manifested photocatalytic activity of twinned $BiPO_4$ was enhanced 8.7 times and 2.5 times compared to pristine $BiPO_4$ for degradation of methyl orange (MO) and acetaminophen (ACE), respectively, while the photocatalytic activity of the obtained multijunctional samples enhanced 35.7 times and 59.5 times compared to twinned $BiPO_4$ for degradation of MO and ACE, respectively.

The oxidation of 2-methylnaphthol (MNL) to 2-methyl-1,4-naphthoquinone (MNQ, menadione, or vitamin K3) with the yield of 78% was accomplished using 30% H_2O_2 and the hydrothermally stable Ti–MMM-2 catalyst (Kholdeeva 2014; Fig. 9).

Also, the photooxidation of 2-allylphenol and phenols bearing alcohol functional groups over Ti–MMM-2 produced p-benzoquinones with good to moderate yields keeping the other oxidizable sites intact (see Fig. 10). For example, 2-hydroxybenzyl alcohol (HBA), which could be alternatively converted to salicylaldehyde, gave 2-hydroxymethyl-p-BQ in 74% yield (Kholdeeva 2014; Bernoud et al. 2015).

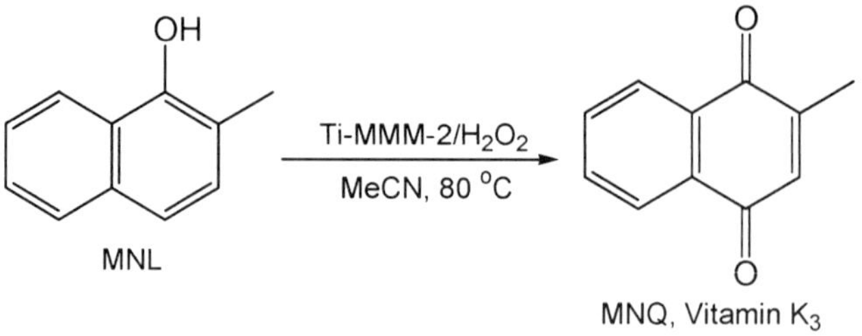

Fig. 9 Scheme revealing the photooxidation of 2-methylnaphthol to 2-methyl-1,4-naphthoquinone. Reproduced with permission from (Kholdeeva 2014). Copyright 2014, Bentham publisher

Fig. 10 Scheme revealing the photooxidation of 2-allylphenol. Reproduced with permission from (Kholdeeva 2014). Copyright 2014, Bentham publisher

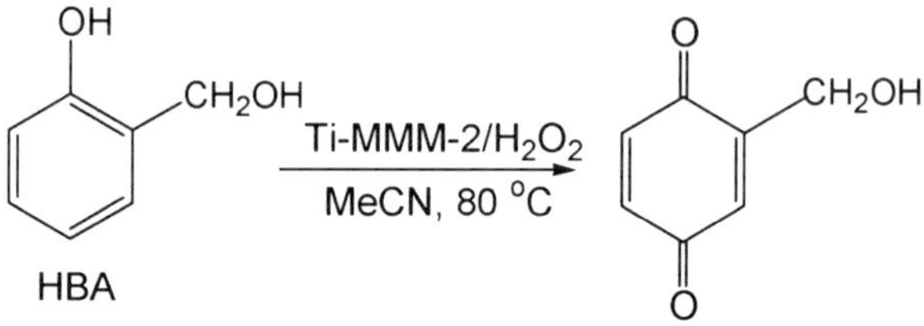

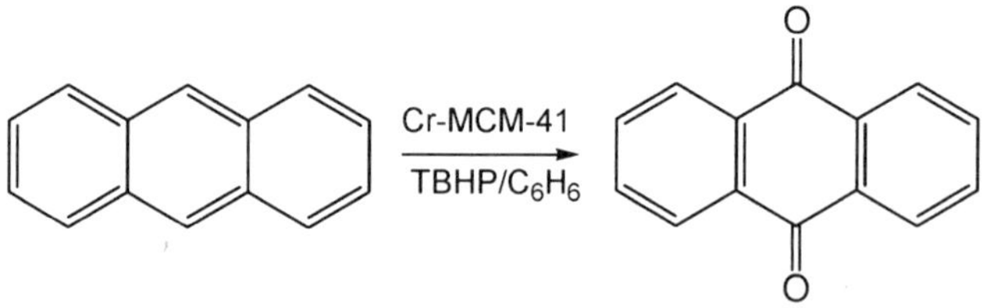

Fig. 11 Scheme revealing the photooxidation of anthracene. Reproduced with permission from (Kholdeeva 2014). Copyright 2014, Bentham publisher

The oxidation of anthracene with tert-butyl hydroperoxide (TBHP) in the present of light was studied using various Ti– M–MCM-41 catalysts (see Fig. 11). Cr–MCM-41 with 0.46 wt% of Cr revealed the best catalytic performance and produced 9,10-anthraquinone (AQ) with ca. 98% selectivity at 79% conversion after 20 h at 80°C (Bernoud et al. 2015; Darkwah et al. 2019).

8 Visible Light-Responsive Nano-based Transition Metal Photocatalysts in Antibacterial Activity

Matsunaga et al. (1985) first reported the bactericidal activity of a UV-responsive photocatalyst. This has since become one of the most researched aspects of transition metals in photocatalysis. Foster et al. (2011) reviewed the photocatalytic disinfection of more than 60 bacterial species, including Gram-positive and Gram-negative bacteria, using a wide range of UV-responsive TiO_2 and other transition metals and substrates. Several groups have conducted studies on the application of impurity-doped visible light-responsive transition metal photocatalysts in bacterial inactivation (Lutic et al. 2012; Mills et al. 1993; Matsunaga et al. 1985; Kau et al. 2009).

Other have also explored possible photocatalytic bactericidal effects against the model bacteria *E. coli* using silver-modified TiO_2 (Gao et al. 2012; Lutic et al. 2012), Ag/N-co-doped TiO_2 (Xu et al. 2008), Ag/C/S co-doped TiO_2 (Hamal and Klabunde 2010), AgI/TiO_2, and Ag/AgBr/TiO_2 visible light-responsive photocatalysts (Hamal and Klabunde 2010). Subramonian and Ta (2014) reported using an N-doped visible light-responsive TiO_2 photocatalyst to reduce the numbers of several human pathogens including *S. flexneri*, *L. monocytogenes*, *V. parahaemolyticus*, *S. aureus*, *S. pyogenes*, and *A. baumannii*. Zhang et al. (2017) investigated the antibacterial activity of the visible light-irradiated C-doped TiO_2 on human pathogens, including *S. flexneri*, *S. aureus*, *and A. baumannii*, observing that photocatalysis was effective against these strains. Among these microorganisms, *S. flexneri*, *L. monocytogenes*, and *V. parahaemolyticus* usually exist in contaminated water, plants, and sewage (Lima et al. 2001) and frequently cause outbreaks in regions with poor public health conditions.

Figure 12 shows visible light-induced bactericidal activities of TiO2-related substrates against *B. subtilis* after illumination at 4uC. It further illustrates that illumination was carried out at a light density of 3610^4 lx (90 mW/cm^2) for either 1 min or 5 min.

The *S. pyogenes* and *S. aureus* species are exotoxin-producing pathogens which can cause many infectious diseases. Visible light-responsive photocatalysts such as Ag/C/S-co-doped TiO2, AgI/TiO2, and Ag/AgBr/TiO2 are well known to kill these pathogens (Hu et al. 2006; Sharma et al. 2016)). Hu et al. (2006) reported using an

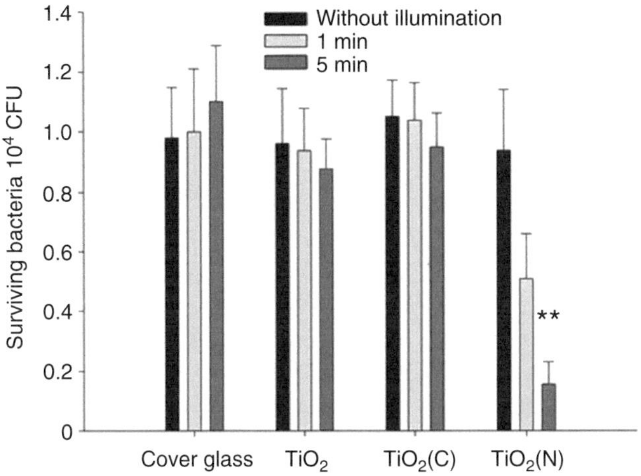

Fig. 12 Bactericidal activity against *B. subtilis*. Data obtained with permission from (Kau et al. 2009). Copyright 2009, PLosOne

N-doped visible light-responsive TiO_2 photocatalyst to reduce the numbers of several human pathogens including *S. flexneri*, *L. monocytogenes*, *V. parahaemolyticus*, *S. aureus*, *S. pyogenes*, and *A. baumannii*. The antibacterial activity of the visible light-irradiated C-doped ZnO on human pathogens, including *S. flexneri*, *S. aureus*, and *A. baumannii*, was analyzed. Results show that photocatalysis was effective against these strains (Zhao et al. 2018). Among these microorganisms, *S. flexneri*, *L. monocytogenes*, and *V. parahaemolyticus* usually exist in contaminated water, plants, and sewage (Lima et al. 2001) and frequently cause outbreaks in regions with poor public health conditions. The *S. pyogenes* and *S. aureus* species are exotoxin-producing pathogens which can cause soft tissue infections, food-borne disease, and toxic shock syndrome (Cheng et al. 2009).

Multidrug-resistant *A. baumannii* causing nosocomial infections are rapid spreading and of great concern in public health (Izdebski et al. 2015). The N-doped FeO_2 and TiO_2 photocatalysts (Fig. 12) were also proven effective against *B. subtilis*, *B. thuringiensis*, *B. cereus*, and *B. anthracis* under visible light illumination (Kau et al. 2009).

To improve the bactericidal performance of the visible light-responsive transition metal photocatalysts, a recent study synthesized silver nanostructure coated N- and C-doped TiO_2, ZnO, and MnO photocatalysts. This was tried on human pathogens *S. pyogenes*, *S. aureus*, and multidrug-resistant *A. baumannii* isolated from hospitals (Lutic et al. 2012).

Chaberny et al. (2003) researched on the application of UV-responsive titania photocatalysts as a suspension in a slurry UV reactor, as a thin film coated on a reactor surface, and as a membrane filter. This suggests the potential use of the visible light-responsive transition metal photocatalysts in a variety of settings to reduce the transmission of pathogens and other pollutants in public environments. Although the optimal conditions of antimicrobial treatments using the visible light-responsive transition metal photocatalysts have yet to be fully established, evidence has confirmed their effectiveness against bacteria, including human pathogens (Lake et al. 2016).

9 Comparison Between Photocatalysts, Antibiotics, and PPCPs

Since the first application of photocatalysts in disinfection, investigators who have conducted extensive research on the bactericidal mechanisms (Fig. 12) of photocatalysis (Kohanski et al. 2007) conclude that bactericidal antibiotics, including aminoglycosides, quinolone, and b-lactams, share a common mechanism for the killing of bacteria, despite targeting different components of different cells. These compounds all converge into a common hydroxyl radical production pathway which induces bacterial cell death (Kohanski et al. 2007). This suggests that a bacterium might become multidrug- and photocatalysis-resistant by finding a way to avoid the

Table 2 Comparison of transition metals, PPCPs, and antibiotics in photocatalysis

Photocatalysts	Transition metals	PPCPs	Antibiotics	References
Photocatalysts in disinfection, research on the bactericidal mechanisms described that bactericidal antibiotics, including aminoglycosides, quinolone, and b-lactams, share a common mechanism for the killing of bacteria, despite targeting different components of different cells	Transition metals such as ZnO, MnO, and TiO_2 treatment, under UV-A illumination indicated that in the presence of UV-A, these metals effectively reduced the number of antibiotic-resistant microbes in suspension	The main effective methods currently used for removing PPCP in conventional wastewater treatment plants are biotransformation or biodegradation by the use of photocatalysis and adsorption on conventional adsorbents, activated carbon is a good example	Multidrug-resistant *A. baumannii* are known to be more susceptible to transitional metal photocatalysis than their related antibiotic-sensitive strain	(See Fig. 8), Kohanski et al. (2007); Xu et al. (2019a, b); Alvarino et al. (2018)
Photocatalysts could be used as complementary to existing disinfection technologies and, thus, facilitate the control of the spread of pathogenic bacteria		Ozonation and activated carbon filtration as the required technologies to reduce the emission of micro-pollutants via WWTP effluents as well as advanced oxidation technology in ferrate (Fe(VI)) treatment technology		Awazu et al. (2008); Yang et al. (2012)

effects of the hydroxyl radical production (Table 2; Fig. 13). However, a comparative study of the bactericidal effects of transition metals against antibiotic-resistant and antibiotic-sensitive bacteria produced results to indicate otherwise (Kohanski et al. 2010).

The aforementioned study applied transition metals such as ZnO, MnO, and TiO_2 treatment, under UV-A illumination, to methicillin-resistant *S. aureus* (Xu et al. 2019a), multidrug-resistant *A. baumannii*, and vancomycin-resistant *E. faecalis*. They compared their findings with those obtained using the same treatment on antibiotic-sensitive strains (Duffy et al. 2004). The results indicated that in the presence of UV-A, these metals effectively reduced the number of antibiotic-resistant microbes in suspension. The methicillin-resistant and antibiotic-sensitive *S. aureus* strains were equally susceptible to transition metal photocatalysis.

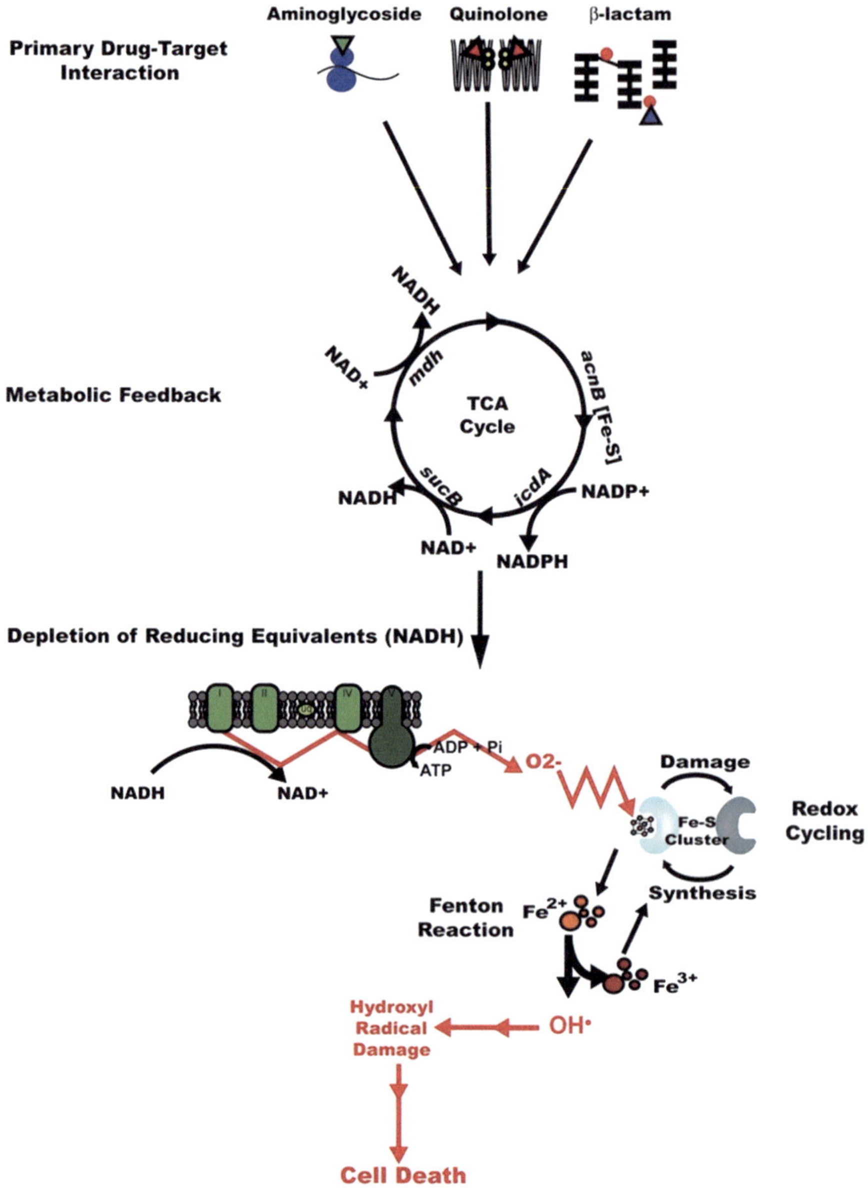

Fig. 13 Proposed model for common mechanism of killing by bactericidal antibiotics Adapted with permission from Kohanski et al. (2007). Copyright 2019, Elsevier

However, multidrug-resistant *A. baumannii* was more susceptible to transitional metal photocatalysis than the antibiotic-sensitive strain. In contrast, vancomycin-resistant *E. faecalis* was less susceptible to the photocatalytic treatment than vancomycin-sensitive *E. faecalis* (Awazu et al. 2008; Xu et al. 2019a). Although the mechanisms for multidrug resistance have yet to be fully elucidated, these results suggest that antibiotic-resistant bacteria are not equally resistant to photocatalysis. This suggests that the use of photocatalysts could be complementary to existing disinfection technologies and, thus, facilitate the control of the spread of pathogenic bacteria.

It was also discovered in a review study carried out by Alvarino et al. (2018) that the main methods currently used for removing PPCPs in conventional wastewater treatment plants are biotransformation or biodegradation and adsorption on conventional adsorbents using activated carbon as a good example (Alvarino et al. 2018). However, conventional wastewater treatment plants are inefficient at removing PPCPs due to their low biodegradability (Li et al. 2017). This therefore requires advanced form of technologies for treatment and complete removal. Yang et al. (2012), however, suggest ozonation and activated carbon filtration as the required technologies to reduce the emission of micro-pollutants via WWTP effluents as well as advanced oxidation technology in ferrate (Fe(VI)) treatment technology.

The primary drug-target interactions (aminoglycoside with the ribosome, quinolone with DNA gyrase, and b-lactam with penicillin-binding proteins) stimulate oxidation of NADH via the electron transport chain that is dependent upon the TCA cycle. Hyperactivation of the electron transport chain stimulates superoxide formation. Superoxide damages iron-sulfur clusters, making ferrous iron available for oxidation by the Fenton reaction. The Fenton reaction leads to hydroxyl radical formation, and the hydroxyl radicals damage DNA, proteins, and lipids, which results in cell death.

Alternative solutions, such as advanced treatment processes, should be implemented to diminish the release of PPCPs into environment. Heterogeneous photocatalysis is considered as a sustainable technique and must be encouraged.

10 Application of Transition Metals in Photocatalysis

The practical and specialized application of transition metals and the synthesis of photocatalytic structures varied in diversified fields. However, the application of photocatalysis in industry could be found in five (5) key areas: ecological purification of industrial wastewaters, industrial gas waste, hydrogen production from different raw materials, CO production in methanol industry, and in solar energy (Ivanenko et al. 2018). Darkwah and Ao (2018), in their review on the structure and properties of photocatalysis, as well as preparation techniques of graphitic carbon nitride nano-based particle, and its applications, outlined some key areas where transition metals could be applied in the photocatalytic processes for the benefit and sustainable environmental management. These include the use of the semiconductor photocatalysts for the removal of ambient intensities of organic and inorganic pollutants from aqueous and gaseous medium in the drinking water treatment, environmental cleanup, and industrial and health applications. They further indicated that the semiconductors, for instance, TiO_2- and ZnO, have enormous ability to oxidize organic and inorganic substances in the air and water through redox processes for its effective use in solar energy utilization, wastewater and environmental treatment, and biomedical and sensing applications without any secondary pollution (Darkwah and Ao 2018).

Transition metals play a key role in the photodegradation process of organic pollutants, so they have become promising materials in the field of photocatalysis. For example, ZnO and TiO_2 have gained a lot of attention due to their various applications in spintronics. They are considered to have low cost and exhibit complete mineralization of organic pollutants when discharged in water (Ramay et al. 2015). Ramay et al. (2015) also inferred that though transition metal application may have a disadvantage of being easily discharged in water which sometimes hinders their recovery and poses a potential loss of materials, there are some solid-liquid separation techniques which may oftentimes come with additional cost. They, therefore, suggested the application of magnetic photocatalyst to be used for recovery and reuse. Recently, photocatalysis studies have revealed that transition metal oxides such as TiO_2 are a possible attractive material because of their potential application in photocatalysis in dye-sensitized solar cells, lithium batteries, and cancer treatment (Shalan et al. 2014).

11 Future Perspectives and Conclusion

To achieve high-quality water supply to society requires a lot of commitments and a strenuous effort by relevant authorities as well as a major responsibility of industry players. Recent years seem to have seen a substantial development in the emergence of effective and efficient processes to remove and degrade various contaminants in wastewater. Modified transition metals are very important in the management and treatment of the environmental pollutants especially those which are used in the photocatalytic processes. This was confirmed by Qi et al. (2020) in their research which indicated that noble metals have significant application prospects in the chemical production processes and environmental protection fields.

The prevalence of pharmaceuticals and personal care products (PPCPs) in aquatic media is mainly due to inefficient removal systems employed by the traditional wastewater treatment plants and also an increase in the pharmaceutical and personal care industries without corresponding sewage treatment facilities to degrade the large numbers of pollutants discharged daily. Numerous techniques for removing PPCPs have been tested, including activated sludge, wetlands, adsorption, and membrane and advanced oxidation processes. Furthermore, the improved adsorption and surface catalysis of the optimized photocatalysts should be ascertained by these transitions modified metals in various photocatalytic experiments such as pollutant degradation and CO_2 reduction (Zada et al. 2019; Xu et al. 2019a; Yasmeen et al. 2019b; Darkwah and Oswald 2019).

Despite the numerous works on this, more research is required to make full use of the excellent photocatalytic and antibacterial performances of transition metal-modified magnetite nanostructured materials. Based on the above discussion, the transition metal-modified nanocomposite can be regarded as a highly efficient photocatalytic and photo-degradative substance for organic pollutant degradation in the future.

Acknowledgements This work was financially supported by grants from the National Science Funds for Creative Research Groups of China (No. 51421006), the National Natural Science Foundation of China (51679063), the Key Program of National Natural Science Foundation of China (No. 41430751), the National Key Plan for Research and Development of China 678 (2016YFC0502203), Fundamental Research Funds (No. 2016B43814), PAPD, National Natural Science Foundation of China (51878316), Jilin Science Foundation for Excellent Young Scholars (20180520169JH), Science and Technology Research Project of Jilin Provincial Department of Education (JJKH20190878KJ), and African Environmental Sanitation (AfES) Consult, Accra, Ghana.

Conflict of Interest Statement The authors declare no conflict of interest.

References

Adormaa BB, Darkwah WK, Ao Y (2018) Oxygen vacancies of the TiO_2 nano-based composite photocatalysts in visible light responsive photocatalysis. RSC Adv 8:33551–33563. https://doi.org/10.1039/C8RA05117H

Akkari M, Aranda P, Belver C, Bedia J, Ben Haj Amara A, Ruiz-Hitzky E (2018) ZnO/sepiolite heterostructured materials for solar photocatalytic degradation of pharmaceuticals in wastewater. Appl Clay Sci 156:104–109. https://doi.org/10.1016/j.clay.2018.01.021

Alexander JT, Hai FI, Alaboud TM (2012) Chemical coagulation-based processes for trace organic contaminant removal: current state and future potential. J Environ Manag 111:195–207

Ali N, Zada A, Zahid M, Ismail A, Rafiq M, Riaz A, Khan A (2018) Enhanced photodegradation of methylene blue with alkaline and transition-metal ferrite nanophotocatalysts under direct sun light irradiation. J Chin Chem Soc:1–7. https://doi.org/10.1002/jccs.201800213

Alvarino T, Suarez S, Lema J, Omil F (2018) Understanding the sorption and biotransformation of organic micropollutants in innovative biological wastewater treatment technologies. Sci Total Environ 615:297–306. https://doi.org/10.1016/j.scitotenv.2017.09.278

Asif MB, Ansari AJ, Chen S, Nghiem LD, Price WE, Hai FI (2018) Understanding the mechanisms of trace organic contaminant removal by high retention membrane bioreactors: a critical review. Environ Sci Pollut Res:1–16. https://doi.org/10.1007/s11356-018-3256-8

Awazu K, Fujimaki M, Rockstuhl C, Tominaga J, Murakami H, Ohki Y, Yoshida N, Watanabe T (2008) A plasmonic photocatalyst consisting of silver nanoparticles embedded in titanium dioxide. J Am Chem Soc 130:1676–1680. https://doi.org/10.1021/ja076503n

Awfa D, Ateia M, Fujii M, Johnson MS, Yoshimura C (2018) Photodegradation of pharmaceuticals and personal care products in water treatment using carbonaceous-TiO2 composites: a critical review of recent literature. Water Res 142:26–45. https://doi.org/10.1016/j.watres.2018.05.036

Bai S, Wang L, Li Z, Xiong Y (2017) Facet-engineered surface and interface design of photocatalytic materials. Adv Sci 4. https://doi.org/10.1002/advs.201600216

Barrocas B, Neves MC, Conceição Oliveira M, Monteiro OC (2018) Enhanced photocatalytic degradation of psychoactive substances using amine-modified elongated titanate nanostructures. Environ Sci Nano 5:350–361. https://doi.org/10.1039/c7en00882a

Benaboud A, Zaabat M, Aida MS, Boudine B, Benzitouni S, Saidani T (2017) Fe2O4/ZnO-nanowires synthesis by dip-coating for Orange II-dye photodegradation. Optik (Stuttg) 144:397–405. https://doi.org/10.1016/j.ijleo.2017.06.108

Bernoud E, Lepori C, Mellah M, Schulz E, Hannedouche J (2015) Recent advances in metal free- and late transition metal-catalysed hydroamination of unactivated alkenes. Cat Sci Technol 5:2017–2037. https://doi.org/10.1039/c4cy01716a

Borges ME, Sierra M, Cuevas E, García RD, Esparza P (2016) Photocatalysis with solar energy: sunlight-responsive photocatalyst based on TiO2loaded on a natural material for wastewater treatment. Sol Energy 135:527–535. https://doi.org/10.1016/j.solener.2016.06.022

Boxall AB et al (2012) Pharmaceuticals and personal care products in the environment: what are the big questions? Environ Health Perspect 120(9):1221–1229

Bu Q, Wang B, Huang J, Deng S (2013) Yu pharmaceuticals and personal care products in the aquatic environment in China: a review. J Hazard Mater 262:189–211

Chaberny IF, Massholder K, Stickler M, Klaus PK (2003) Disinfection of surfaces by photocatalytic oxidation with titanium dioxide and UVA light. Chemosphere 53:71–77. https://doi.org/10.1016/S0045-6535(03)00362-X

Chen KC, Wang YH, Lu YC (2011) Treatment of polluted water for reclamation using photocatalysis and constructed wetlands. Catal Today 175:276–282. https://doi.org/10.1016/j.cattod.2011.06.013

Chen F, Gong Z, Kelly BC (2017) Bioaccumulation behavior of pharmaceuticals and personal care products in adult Zebrafish (Danio rerio): influence of physical-chemical properties and biotransformation. Environ Sci Technol 51(19):11085–11095

Chen P, Zhang Q, Shen L, Li R, Tan C, Chen T, Liu H, Liu Y, Cai Z, Liu G, Lv W (2019) Insights into the synergetic mechanism of a combined vis-RGO/TiO2/peroxodisulfate system for the degradation of PPCPs: kinetics, environmental factors and products. Chemosphere 216:341–351. https://doi.org/10.1016/j.chemosphere.2018.10.096

Cheng C, Sun D, Chu W, Tseng Y, Ho H, Wang J, Chung P, Chen J, Tsai J, Lin N, Yu M, Chang H (2009) The effects of the bacterial interaction with visible-light responsive titania photocatalyst on the bactericidal performance. J Biomed Sci 16(7):1–10. https://doi.org/10.1186/1423-0127-16-7

Cizmas L, Sharma VK, Gray CM, McDonald TJ (2015) Pharmaceuticals and personal care products in waters: occurrence, toxicity, and risk. Environ Chem Lett 13(4):381–394

Czech B, Tyszczuk-Rotko K (2018) Visible-light-driven photocatalytic removal of acetaminophen from water using a novel MWCNT-TiO2-SiO2 photocatalysts. Sep Purif Technol 206:343–355. https://doi.org/10.1016/j.seppur.2018.06.025

Darkwah WK, Ao Y (2018) Mini review on the structure and properties (Photocatalysis), and preparation techniques of graphitic carbon nitride Nano-based particle, and its applications. Nanoscale Res Lett 13. https://doi.org/10.1186/s11671-018-2702-3

Darkwah WK, Oswald KA (2019) Photocatalytic applications of Heterostructure graphitic carbon nitride: pollutant degradation, hydrogen gas production, and CO_2 reduction. Nanoscale Res Lett 14:234. https://doi.org/10.1186/s11671-019-3070-3

Darkwah WK, Adormaa BB, Christelle Sandrine MK, Ao Y (2019) Modification strategies for enhancing the visible light responsive photocatalytic activity of the BiPO 4 nano-based composite photocatalysts. Cat Sci Technol 9:546–566. https://doi.org/10.1039/c8cy02039f

Darkwah WK, Teye GK, AoY (2020) Nanoparticles - graphene nanocrystals in CO2 photoreduction with H2O for fuel production. Nanoscale Adv 2:991–1006. https://doi.org/10.1039/C9NA00756C

Dellarocas C, Becerra-Fernandez I, Boord L (2009) United States patent 7509230. Method for rating an entity. United States Patent Office

Deng Y, Zhao R (2015) Advanced oxidation processes (AOPs) in wastewater treatment. Curr Pollut Rep 1:167–176

Deng X, Zhu L, Zhang H, Kroner A, Zheng J, Zhang N, He J, Hui B (2018) ScienceDirect ruthenium stabilized on transition metal-on- transition metal oxide nanoparticles for naphthalene hydrogenation. Int J Hydrogen Energy:1–9. https://doi.org/10.1016/j.ijhydene.2018.05.170

Djouadi L, Khalaf H, Boukhatem H, Boutoumi H, Kezzime A, Santaballa JA, Canle M (2018) Degradation of aqueous ketoprofen by heterogeneous photocatalysis using Bi 2 S 3 /TiO 2 – Montmorillonite nanocomposites under simulated solar irradiation. Appl Clay Sci 166:27–37. https://doi.org/10.1016/j.clay.2018.09.008

Duffy EF, Al Touati F, Kehoe SC, Mcloughlin OA, Gill LW, Gernjak W, Oller I, Maldonado MI, Malato S, Cassidy J (2004) A novel TiO 2 -assisted solar photocatalytic batch-process disinfection reactor for the treatment of biological and chemical contaminants in domestic drinking water in developing countries. Sol Energy 77(5):649–655. https://doi.org/10.1016/j.solener.2004.05.006

Durairaj A, Sakthivel T, Obadiah A, Vasanthkumar S (2018) Enhanced photocatalytic activity of transition metal ions doped g–C3N4 nanosheet activated by PMS for organic pollutant degradation. J Mater Sci Mater Electron 29:8201–8209. https://doi.org/10.1007/s10854-018-8826-5

Ebele AJ, Abdallahab MA-E, Harrad S (2017) Pharmaceuticals and personal care products (PPCPs) in the freshwater aquatic environment. Emerg Contam 3:1–16

Esplugas S, Bila DM, Krause LGT, Dezotti M (2007) Ozonation and advanced oxidation technologies to remove endocrine disrupting chemicals (EDCs) and pharmaceuticals and personal care products (PPCPs) in water effluents. J Hazard Mater 149:631–642. https://doi.org/10.1016/j.jhazmat.2007.07.073

Fernández RL, McDonald JA, Khan SJ, Le-Clech P (2014) Removal of pharmaceuticals and endocrine disrupting chemicals by a submerged membrane photocatalysis reactor (MPR). Sep Purif Technol 127:131–139. https://doi.org/10.1016/j.seppur.2014.02.031

Fernández-Ibáñeza P, Sichel P, Polo-López MI, de Cara-García M, Tello JC (2009) Photocatalytic disinfection of natural well water contaminated by Fusarium solani using TiO 2 slurry in solar CPC photo-reactors. Catal Today 144(1):62–68. https://doi.org/10.1016/j.cattod.2009.01.039

Foster HA, Ditta IB, Varghese S, Steele A (2011) Photocatalytic disinfection using titanium dioxide: spectrum and mechanism of antimicrobial activity. Appl Microbiol Biotechnol 90:1847–1868. https://doi.org/10.1007/s00253-011-3213-7

Freyria FS (2018) Nanomaterials for the abatement of pharmaceuticals and personal care products from wastewater. Appl Sci 8(2):170. https://doi.org/10.3390/app8020170

Fujishima A, Nakata K, Ochiai T, Manivannan A, Tryk DA (2013) Recent aspects of photocatalytic technologies for solar fuels, self-cleaning, and environmental cleanup. Interface Mag 22:51–56. https://doi.org/10.1149/2.F06132if

Fung CSL, Khan M, Kumar A, Lo IMC (2019) Visible-light-driven photocatalytic removal of PPCPs using magnetically separable bismuth oxybromo-iodide solid solutions: mechanisms, pathways, and reusability in real sewage. Sep Purif Technol 216:102–114. https://doi.org/10.1016/j.seppur.2019.01.077

Gao L, Ren W, Xu H, Jin L, Wang Z, Ma T, Ma L, Zhang Z, Fu Q, Peng L, Bao X, Cheng H (2012) Repeated growth and bubbling transfer of graphene with millimetre-size single-crystal grains using platinum. Nat Commun 3:697–699. https://doi.org/10.1038/ncomms1702

Gaya UI, Abdullah AH, Zainal Z, Hussein MZ (2009) Photocatalytic treatment of 4-chlorophenol in aqueous ZnO suspensions: intermediates, influence of dosage and inorganic anions. J Hazard Mater 168:57–63. https://doi.org/10.1016/j.jhazmat.2009.01.130

Gonzalez-Zavala F, Escobar-Alarcón L, Solís-Casados DA, Rivera-Rodríguez C, Basurto R, Haro-Poniatowski E (2016) Preparation of vanadium oxide thin films modified with Ag using a hybrid deposition configuration. Appl Phys A Mater Sci Process 122:1–6. https://doi.org/10.1007/s00339-016-9991-0

Guo Y, Wang P, Qian J, Hou J, Ao Y, Wang C (2018) Construction of a composite photocatalyst with significantly enhanced photocatalytic performance through combination of homo-junction with hetero-junction. Cat Sci Technol 8:486–498. https://doi.org/10.1039/c7cy02027a

Habashi F, City Q (2017) Transition metals. 10–12. https://doi.org/10.1016/B978-0-12-803581-8.10371-6

Hai FI, Nghiem LD, Khan SJ, Price WE, Yamamoto K (2014) Wastewater reuse: removal of emerging trace organic contaminants. In: Hai FI, Yamamoto K, Lee CH (eds) Membrane biological reactors. IWA Publishing, London

Hai FI, Yang S, Asif MB, Sencadas V, Shawkat S, Sanderson-Smith M, Gorman J, Xu Z-Q, Yamamoto K (2018) Carbamazepine as a possible anthropogenic marker in water: occurrences, toxicological effects, regulations and Remova by wastewater treatment technologies. Water 10:107

Hamal DB, Klabunde KJ (2010) Heterogeneous photocatalysis over silver halide photocatalysts for environmental remediation. ACS Symp Ser 1045:191–205. https://doi.org/10.1021/bk-2010-1045.ch011

Hermannsdörfer J, Friedrich M, Miyajima N, Albuquerque RQ, Kümmel S, Kempe R (2012) Ni/Pd@MIL-101: synergistic catalysis with cavity-conform Ni/Pd nanoparticles. Angew Chem Int Ed 51:11473–11477. https://doi.org/10.1002/anie.201205078

Hu C, Hu X, Guo J, Qu J (2006) Efficient destruction of pathogenic bacteria with NiO/SrBi2O4 under visible light irradiation. Environ Sci Technol 40:5508–5513

Ivanenko I, Dontsova T, Fedenko Y (2018) Applications perspectives of nanodispersed chalcogenides of transition metals in photocatalysis. Springer Proc Phys 214:99–113. https://doi.org/10.1007/978-3-319-92567-7_7

Izdebski R, Fiett J, Urbanowicz P, Baraniak A, Derde LPG, Bonten MJM, Carmeli Y, Goossens H, Hryniewicz W, Brisse S, Gniadkowski M (2015) Phylogenetic lineages, clones and b -lactamases in an international collection of Klebsiella oxytoca isolates non-susceptible to expanded-spectrum cephalosporins. J Antimicrob Chemother 70(12):3230–3237. https://doi.org/10.1093/jac/dkv273

Jiang JQ, Zhou Z, Sharma VK (2013) Occurrence, transportation, monitoring and treatment of emerging micro-pollutants in waste water – a review from global views. Microchem J 110:292–300. https://doi.org/10.1016/j.microc.2013.04.014

Kagle J, Porter AW, Murdoch RW, Rivera-Cancel G, Hay AG (2009) Biodegradation of pharmaceutical and personal care products. In: Advances in applied microbiology, 1st edn. Elsevier, Amsterdam. https://doi.org/10.1016/S0065-2164(08)01003-4

Kau J, Sun D, Huang H, Wong M, Lin H (2009) Role of visible light-activated photocatalyst on the reduction of Anthrax spore-induced mortality in mice. PLoS One 4(1):4167. https://doi.org/10.1371/journal.pone.0004167

Khan MM, Pradhan D, Sohn Y (2017) Nanocomposites for visible light-induced photocatalysis. Springer series on polymer and composite materials. https://doi.org/10.1007/978-3-319-62446-4

Kholdeeva OA (2014) Recent developments in liquid-phase selective oxidation using environmentally benign oxidants and mesoporous metal silicates. Cat Sci Technol 4:1869–1889. https://doi.org/10.1039/c4cy00087k

Khraisheh M, Kim J, Campos L, Al-muhtaseb AH, Al-hawari A, Al M, Walker GM (2014) Journal of industrial and engineering chemistry removal of pharmaceutical and personal care products (PPCPs) pollutants from water by novel TiO 2 – coconut Shell powder (TCNSP) composite. J Ind Eng Chem 20:979–987. https://doi.org/10.1016/j.jiec.2013.06.032

Kohanski MA, Dwyer DJ, Hayete B, Lawrence CA (2007) A common mechanism of cellular death induced by bactericidal antibiotics. Cell 130(5):97–810. https://doi.org/10.1016/j.cell.2007.06.049

Kohanski MA, Depristo MA, Collins JJ (2010) Article sublethal antibiotic treatment leads to multidrug. Mol Cell 37:311–320. https://doi.org/10.1016/j.molcel.2010.01.003

Lake C, Wu Y, Mei CY, Liu YS (2016) Occurrence of selected PPCPs and sulfonamide resistance genes associated with heavy metals pollution in surface sediments. Environ Earth Sci. https://doi.org/10.1007/s12665-015-4838-0

Leong S, Razmjou A, Wang K, Hapgood K, Zhang X, Wang H (2014) TiO2based photocatalytic membranes: a review. J Membr Sci 472:167–184. https://doi.org/10.1016/j.memsci.2014.08.016

Lester Y, Mamane H, Zucker I, Avisar D (2013) Treating wastewater from a pharmaceutical formulation facility by biological process and ozone. Water Res 47:4349–4356. https://doi.org/10.1016/j.watres.2013.04.059

Li J, Cheng W, Xu L, Strong PJ, Chen H (2015) Antibiotic-resistant genes and antibiotic-resistant bacteria in the effluent of urban residential areas, hospitals, and a municipal wastewater treatment plant system. Environ Sci Pollut Res 22:4587–4596. https://doi.org/10.1007/s11356-014-3665-2

Li J, Zhou Q, Campos LC (2017) Removal of selected emerging PPCP compounds using greater duckweed (Spirodela polyrhiza) based lab-scale free water constructed wetland. Water Res 126:252–261. https://doi.org/10.1016/j.watres.2017.09.002

Liang R, Van Leuwen JC, Bragg LM, Arlos MJ, Li Chun Fong LCM, Schneider OM, Jaciw-Zurakowsky I, Fattahi A, Rathod S, Peng P, Servos MR, Zhou YN (2019) Utilizing UV-LED pulse width modulation on TiO2 advanced oxidation processes to enhance the decomposition efficiency of pharmaceutical micropollutants. Chem Eng J 361:439–449. https://doi.org/10.1016/j.cej.2018.12.065

Lima SAM, Sigoli FA Jr, Jafelicci M, Davolos MR (2001) Luminescent properties and lattice defects correlation on zinc oxide. Int J Inorg Mater 3:749–754

Linsebigler AL, Lu G, Yates JT Jr (1995) Photocatalysis on TiO2 surfaces: principles, mechanisms, and selected results. Chem Rev 95:735–758

Luo Y, Guo W, Ngo HH, Nghiem LD, Hai FI, Zhang J, Liang S, Wang XC (2014) A review on the occurrence of micropollutants in the aquatic environment and their fate and removal during wastewater treatment. Sci Total Environ 473–474:619–641. https://doi.org/10.1016/j.scitotenv.2013.12.065

Lutic D, Coromelci-Pastravanu C, Cretescu I, Poulios I, Stan CD (2012) Photocatalytic treatment of rhodamine 6G in wastewater using photoactive ZnO. Int J Photoenergy 2012. https://doi.org/10.1155/2012/475131

Mathis JE, Lieffers JJ, Mitra C, Reboredo FA, Bi Z, Bridges CA, Kidder MK, Paranthaman MP (2016) Increased photocatalytic activity of TiO2 mesoporous microspheres from codoping with transition metals and nitrogen. Ceram Int 42:3556–3562. https://doi.org/10.1016/j.ceramint.2015.10.164

Matsunaga T, Tomoda R, Nakajima T, Wake H (1985) Photoelectrochemical sterilization of microbial cells by semiconductor powders. FEMS Microbiol Lett 29:211–214

Mc Cullagh JJP, Setchell DJ, Gulabivala K, Hussey DL, Biagioni P, Lamey P, Bailey G (2000) A comparison of thermocouple and infrared thermographic analysis of temperature rise on the root surface during the continuous wave of condensation technique. Int Endod J 33:326–332

Mcloughlin OA, Kehoe SC, Mcguigan KG, Duffy EF, Al Touati F, Gernjak W, Alberola IO, Gill LW, Rodrı SM (2004) Solar disinfection of contaminated water: a comparison of three small-scale reactors. Sol Energy 77(5):657–664. https://doi.org/10.1016/j.solener.2004.07.004

Miklos DB, Remy C, Jekel M, Linden KG, Drewes JE, Hübner U (2018) Evaluation of advanced oxidation processes for water and wastewater treatment – a critical review. Water Res 139:118–131

Miller EL, Nason SL, Karthikeyan KG, Pedersen JA (2015) Root uptake of pharmaceutical and personal care product ingredients. Environ Sci Technol 50(2):525–541. https://doi.org/10.1021/acs.est.5b01546

Mills A, Davies RH, Worsley D (1993) Water purification by semiconductor photocatalysis. Chem Soc Rev 22:417. https://doi.org/10.1039/cs9932200417

Muir D, Simmons D, Wang X, Peart T, Villella M, Jason Miller J, Sherry J (2017) Bioaccumulation of pharmaceuticals and personal care product chemicals in fish exposed to wastewater effluent in an urban wetland. Sci Rep 7:16999

Najm I, Trussel RR (1999) New and emerging drinking water treatment technologies, in identifying future drinking water contaminants. National Academy Press, Washington, pp 33–42

Navntoft C, Araujo P, Litter MI, Apella MC, Fernández D, Puchulu ME, del Hidalgo MV, Blesa MA (2007) Field tests of the solar water detoxification SOLWATER reactor in los Pereyra, Tucumán, Argentina. J Sol Energy Eng 129:127–134

Niu F, Chen D, Qin L, Gao T, Zhang N, Wang S, Chen Z, Wang J, Sun X, Huang Y (2015a) Synthesis of Pt/BiFeO<inf>3</inf> heterostructured photocatalysts for highly efficient visible-light photocatalytic performances. Sol Energy Mater Sol Cells 143:386–396. https://doi.org/10.1016/j.solmat.2015.07.008

Niu F, Chen D, Qin L, Zhang N, Wang J, Chen Z, Huang Y (2015b) Facile synthesis of highly efficient p-n heterojunction CuO/BiFeO3 composite photocatalysts with enhanced visible-light photocatalytic activity. ChemCatChem 7:3279–3289. https://doi.org/10.1002/cctc.201500634

O'Connor T, Panov MS, Mereshchenko A, Tarnovsky AN, Lorek R, Perera D, Diederich G, Lambright S, Moroz P, Zamkov M (2012) The effect of the charge-separating interface on exciton dynamics in photocatalytic colloidal heteronanocrystals. ACS Nano 6:8156–8165. https://doi.org/10.1021/nn302810y

Qi K, Xing X, Zada A, Li M, Wang Q, Liu SY, Lin H, Wang G (2020) Transition metal doped ZnO nanoparticles with enhanced photocatalytic and antibacterial performances: experimental and DFT studies. Ceram Int 46:1494–1502. https://doi.org/10.1016/j.ceramint.2019.09.116

Qin Z, Wang M, Li R, Chen Y (2018) Novel Cu3P/g-C3N4 p-n heterojunction photocatalysts for solar hydrogen generation. Sci China Mater 61:861–868. https://doi.org/10.1007/s40843-017-9171-9

Radjenović J, Petrović M, Barceló D (2009) Fate and distribution of pharmaceuticals in wastewater and sewage sludge of the conventional activated sludge (CAS) and advanced membrane bioreactor (MBR) treatment. Water Res 43:831–841

Ramay SM, Mahmood A, Atiq S, Siddiqi SA, Naseem S (2015) Photodegradation of organic pollutants using Fe-doped ZnO, materials today: proceedings. Elsevier, Amsterdam. https://doi.org/10.1016/j.matpr.2015.11.074

Reemtsma T, Berger U, Arp HPH, Gallard H, Knepper TP, Neumann M, Quintana JB, de Voogt P (2016) Mind the gap: persistent and Mobile organic compounds – water contaminants that slip through. Environ Sci Technol 50:10308–10315

Richardson BJ, Lam PK, Martin M (2005) Emerging chemicals of concern: pharmaceuticals and personal care products (PPCPs) in Asia, with particular reference to southern China. Mar Pollut Bull 50(9):913–920

Savun-Hekimoğlu B, Ince NH (2017) Decomposition of PPCPs by ultrasound-assisted advanced Fenton reaction: a case study with salicylic acid. Ultrason Sonochem 39:243–249. https://doi.org/10.1016/j.ultsonch.2017.04.013

Schwarzenbach RP, Escher BI, Fenner K, Hofstetter TB, Johnson CA, von Gunten U, Wehrli B (2006) The challenge of micro pollutants in aquatic systems. Science 313:1072–1077

Schymanski EL, Singer HP, Longree P, Loos M, Ruff M, Stravs MA, Ripolles Vidal C, Hollender J (2014) Strategies to characterize polar organic contamination in wastewater: exploring the capability of high resolution mass spectrometry. Environ Sci Technol 48:1811–1818

Shalan AE, Rasly M, Rashad MM (2014) Organic acid precursor synthesis and environmental photocatalysis applications of mesoporous anatase TiO2 doped with different transition metal ions. J Mater Sci Mater Electron 25:3141–3146. https://doi.org/10.1007/s10854-014-1995-y

Sharma VK, Johnson N, Cizmas L, McDonald TJ, Kim H (2016) A review of the influence of treatment strategies on antibiotic resistant bacteria and antibiotic resistance genes. Chemosphere 150:702–714. https://doi.org/10.1016/j.chemosphere.2015.12.084

Shi JW, Xie C, He C, Liu C, Gao C, Yang S, Chen JW, Li G (2015) Photocatalytic performance comparison of titania hollow spheres composed of nanoplates with dominant {001} facets and nanoparticles without dominant {001} facets. Catal Commun 66:46–49. https://doi.org/10.1016/j.catcom.2015.03.016

Shon HK, Phuntsho S, Vigneswaran S (2008) Effect of photocatalysis on the membrane hybrid system for wastewater treatment. Desalination 225:235–248. https://doi.org/10.1016/j.desal.2007.05.032

Silva CG, Monteiro J, Marques RRN, Silva AMT, Martínez C, Canle LM, Faria JL (2013) Photochemical and photocatalytic degradation of trans-resveratrol. Photochem Photobiol Sci 12:638–644. https://doi.org/10.1039/c2pp25239b

Strohfeldt KA (2015) Transition metals and d-block metal chemistry. Essentials Inorg Chem:123–182. https://doi.org/10.1002/9781118695425.ch7

Subramonian W, Ta Y (2014) Effect of enhancers and inhibitors on photocatalytic sunlight treatment of methylene blue. Water Air Soil Pollut 225(4). https://doi.org/10.1007/s11270-014-1922-0

Tang CY, Yang Z, Guo H, Wen JJ, Nghiem LD, Cornelissen E (2018) Potable water reuse through advanced membrane technology. Environ Sci Technol 52(18). https://doi.org/10.1021/acs.est.8b00562

Teixeira B, Pinto T, Silva F, Santos G, Praça I, Vale Z (2018) Multi-agent decision support tool to enable interoperability among heterogeneous energy systems. Appl Sci 8:328. https://doi.org/10.3390/app8030328

Tijani JO, Fatoba OO, Babajide OO, Petrik LF (2016) Pharmaceuticals, endocrine disruptors, personal care products, nanomaterials and perfluorinated pollutants: a review. Environ Chem Lett 14:27–49. https://doi.org/10.1007/s10311-015-0537-z

Tran NH, Reinhard M, Gin KYH (2018) Occurrence and fate of emerging contaminants in municipal wastewater treatment plants from different geographical regions – a review. Water Res 133:182–207

Villén L, Manjón F, García-Fresnadillo D, Orellana G (2006) Solar water disinfection by photocatalytic singlet oxygen production in heterogeneous medium. Appl Catal B Environ 69:1–9. https://doi.org/10.1016/j.apcatb.2006.05.015

Wang J, Wang S (2016a) Removal of pharmaceuticals and personal care products (PPCPs) from wastewater: a review. J Environ Manag 182:620–640. https://doi.org/10.1016/j.jenvman.2016.07.049

Wang J, Wang S (2016b) Removal of pharmaceuticals and personal care products (PPCPs) from wastewater: a review. J Environ Manage 182:620–640. https://doi.org/10.1016/j.jenvman.2016.07.049

Wang Y, Wang X, Li M, Dong J, Sun C, Chen G (2018a) Removal of pharmaceutical and personal care products (PPCPs) from municipal waste water with integrated membrane systems, MBR-RO / NF. Int J Environ Res Public Health 15:269. https://doi.org/10.3390/ijerph15020269

Wang H, Zhou P, Guo R, Wang Y, Zhan H, Yuan Y (2018b) Synthesis of Rectorite/Fe3O4/ZnO composites and their application for the removal of methylene blue dye. Catalysts 8:107. https://doi.org/10.3390/catal8030107

Xu H, Li H, Wu C, Chu J, Yan Y, Shu H (2008) Preparation, characterization and photocatalytic activity of transition metal-loaded BiVO4. Mater Sci Eng B Solid-State Mater Adv Technol 147:52–56. https://doi.org/10.1016/j.mseb.2007.11.011

Xu T, Zheng H, Zhang P, Lin W, Sekiguchi Y (2015) Hydrothermal preparation of nanoporous TiO<inf>2</inf> films with exposed {001} facets and superior photocatalytic activity. J Mater Chem A 3:19115–19122. https://doi.org/10.1039/c5ta02640g

Xu C, Ravi Anusuyadevi P, Aymonier C, Luque R, Marre S (2019a) Nanostructured materials for photocatalysis. Chem Soc Rev 48(14):3868–3902. https://doi.org/10.1039/c9cs00102f

Xu B, Zada A, Wang G, Qu Y (2019b) Boosting the visible-light photoactivities of BiVO4 nanoplates by doping Eu and coupling CeOx nanoparticles for CO2 reduction and organic oxidation. Sustain Energy Fuels 3:3363–3369. https://doi.org/10.1039/C9SE00409B

Yang B, Ying G, Zhao J, Liu S, Zhou L, Chen F (2012) Removal of selected endocrine disrupting chemicals (EDCs) and pharmaceuticals and personal care products (PPCPs) during ferrate (VI) treatment of secondary wastewater effluents. Water Res 46:2194–2204. https://doi.org/10.1016/j.watres.2012.01.047

Yasmeen H, Zada A, Li W, Xu M, Liu S (2019a) Suitable energy platform of Bi2WO6 significantly improves visible-light degradation activity of g-C3N4 for highly toxic diuron pollutant. Mater Sci Semicond Process 102:104598. https://doi.org/10.1016/j.mssp.2019.104598

Yasmeen H, Zada A, Liu S (2019b) Dye loaded MnO 2 and chlorine intercalated g-C 3 N 4 coupling impart enhanced visible light photoactivities for pollutants degradation. J Photochem Photobiol A Chem 380:111867. https://doi.org/10.1016/j.jphotochem.2019.111867

Yi C, Liao Q, Deng W, Huang Y, Mao J, Zhang B, Wu G (2019) The preparation of amorphous TiO2 doped with cationic S and its application to the degradation of DCFs under visible light irradiation. Sci Total Environ 684:527–536. https://doi.org/10.1016/j.scitotenv.2019.05.338

Yu S, Liu J, Zhou Y, Webster RD, Yan X (2017) Effect of synthesis method on the nanostructure and solar-driven photocatalytic properties of TiO2-CuS composites. ACS Sustain Chem Eng 5:1347–1357. https://doi.org/10.1021/acssuschemeng.6b01769

Zada A, Ali N, Subhan F, Anwar N, Ali Shah MI, Ateeq M, Hussain Z, Zaman K, Khan M (2019) Suitable energy platform significantly improves charge separation of g-C3N4 for CO2 reduction and pollutant oxidation under visible-light. Prog Nat Sci Mater Int 29:138–144. https://doi.org/10.1016/j.pnsc.2019.03.004

Zeng Y, Chen D, Chen T, Cai M, Zhang Q, Xie Z, Li R, Xiao Z, Liu G, Lv W (2019) Study on heterogeneous photocatalytic ozonation degradation of ciprofloxacin by TiO 2 /carbon dots: kinetic, mechanism and pathway investigation. Chemosphere 227:198–206. https://doi.org/10.1016/j.chemosphere.2019.04.039

Zhang G, Wu B, Zhang S (2017) Effects of acetylacetone on the photoconversion of pharmaceuticals in natural and pure waters. Environ Pollut 225:691–699. https://doi.org/10.1016/j.envpol.2017.01.089

Zhang Q, Tan C, Zheng X, Chen P, Zhuo M, Chen T, Xie Z, Wang F, Liu H, Liu Y, Zhang X, Lv W, Liu G (2019) Dual metal-free polymer reactive sites for the efficient degradation of diclofenac by visible light-driven oxygen reduction to superoxide radical and hydrogen peroxide. Environ Sci Nano 6:2577–2590. https://doi.org/10.1039/C9EN00482C

Zhao L, Deng J, Sun P, Liu J, Ji Y, Nakada N, Qiao Z, Tanaka H, Yang Y (2018) Nanomaterials for treating emerging contaminants in water by adsorption and photocatalysis: systematic review and bibliometric analysis. Sci Total Environ 627:1253–1263. https://doi.org/10.1016/j.scitotenv.2018.02.006

Zhao GY, Li CY, Xu Y (2019) PPECu/NiFe2O4 as an efficient visible-light-driven difunctional photocatalyst for degradation of PPCPs and hydrogen production. J Alloys Compd 780:534–539. https://doi.org/10.1016/j.jallcom.2018.11.292

Zhou SL, Zhang S, Liu F, Liu JJ, Xue JJ, Yang DJ, Chang CT (2016) ZnO nanoflowers photocatalysis of norfloxacin: effect of triangular silver nanoplates and water matrix on degradation rates. J Photochem Photobiol A Chem 328:97–104. https://doi.org/10.1016/j.jphotochem.2016.03.037

Zhou Z, Zhang Y, Shen Y, Liu S, Zhang Y (2018) Molecular engineering of polymeric carbon nitride: advancing applications from photocatalysis to biosensing and more. Chem Soc Rev 47:2298–2321. https://doi.org/10.1039/c7cs00840f

Zhu S, Wang D (2017) Photocatalysis: basic principles, diverse forms of implementations and emerging scientific opportunities. Adv Energy Mater 1700841:1700841. https://doi.org/10.1002/aenm.201700841

Effects of Carbamazepine in Bivalves: A Review

Ângela Almeida, Valdemar I. Esteves, Amadeu M. V. M. Soares, and Rosa Freitas

Contents

Abstract Carbamazepine (CBZ) is among the ten most frequent pharmaceuticals that occur in the aquatic systems, with known effects on inhabiting organisms, including bivalves. Bivalves are important species in coastal ecosystems, often exhibiting a dominant biomass within invertebrate communities. These organisms play a major role in the functioning of the ecosystem and particularly in food webs (as suspension-feeders) and represent a significant fraction of the fisheries resource. They also have strong interactions with the environment, water and sediment and are considered good bioindicator species. The present paper reviews the known literature on the impacts of CBZ in biological endpoints of marine bivalves exposed to environmentally and non-environmentally relevant concentrations, highlighting differences in terms of biological responses, associated with exposure period, concentrations tested, and species used. Overall, the literature available showed that CBZ induces individual and sub-individual effects in marine bivalves (adults and life stages) and the most common effect reported was the induction of oxidative stress.

Â. Almeida · A. M. V. M. Soares · R. Freitas (✉)
Biology Department and CESAM, University of Aveiro, Aveiro, Portugal
e-mail: aaalmeida@ua.pt; asoares@ua.pt; rosafreitas@ua.pt

V. I. Esteves
Chemistry Department and CESAM, University of Aveiro, Aveiro, Portugal
e-mail: valdemar@ua.pt

P. de Voogt (ed.), *Reviews of Environmental Contamination and Toxicology Volume 254*, Reviews of Environmental Contamination and Toxicology 254,
https://doi.org/10.1007/398_2020_51

Keywords Antiepileptic · Bivalves · Marine environment · Pharmaceutical drugs · Sub-lethal effects

1 Introduction

Aquatic systems are increasingly exposed to contaminants, mainly derived from human activities, representing the ultimate sink areas for most of the substances (Sousa et al. 2018, among others). Among these contaminants are pharmaceuticals (Monteiro and Boxall 2010). After administration, pharmaceuticals are excreted in urine or feces as a mixture of metabolites, as the unchanged compound, or conjugated with an inactivating substituent (Halling-Sørensen et al. 1998). These substances can reach the aquatic systems through sewage from WasteWater Treatment Plants (WWTPs) as several pharmaceuticals have shown to resist conventional sewage treatment (Monteiro and Boxall 2010). Since drugs are designed to exert biological effects, when in aquatic systems, they can interact with wildlife as they keep their lipophilic capacity to pass cellular membranes. Thus, under environmental conditions, these compounds may exhibit the potential to be bioaccumulated and exert biological effects on non-target organisms (Halling-Sørensen et al. 1998). In fact, negative effects on aquatic organisms have been related with the presence of pharmaceutical drugs in water bodies. A growing literature is available on the evaluation of the effects of these stressors on subcellular and cellular processes in freshwater and marine organisms. To assess the effects, biomarkers that measure physiological, morphological, biochemical, and molecular changes have been applied (Brausch et al. 2012; Fabbri and Franzellitti 2016). In marine organisms (from early life stages to adulthood) submitted to pharmaceutical drugs, the toxicity tests reported acute and chronic effects with individual and sub-individual responses (for reviews see Fabbri and Franzellitti (2016) and Prichard and Granek (2016)). Marine bivalves (e.g., mussels, clams, oysters) are examples of organisms with high ecological and socioeconomic value (Voultsiadou et al. 2010) in which pharmaceutical drugs have been detected. These organisms are of interest due to their socio-economic value, as they are highly consumed shellfish (Voultsiadou et al. 2010). However, the ingestion of bivalves contaminated with drugs may pose a health risk to the human population (e.g., potential antibiotic resistance) (Rodrigues et al. 2019). Moreover, bivalves have been commonly used as bioindicator species as they exhibit a wide distribution, bioaccumulation potential, well-known life cycles, a sedentary lifestyle, and filter feeding activity that suit their use in ecotoxicology and monitoring programs (McEneff et al. 2014; Powell and Cummins 1985; Viarengo and Canesi 1991). Concerning the use of marine bivalves in toxicological studies, in general, at the individual level, pharmaceuticals induced, for example, changes in growth, reproduction, and behavior (e.g., feeding) (e.g., Honkoop et al. 1999; Solé et al. 2010). At the sub-individual level, pharmaceutical drugs led to changes in physiological, cellular, and molecular mechanisms. For example, alteration of immunological, antioxidant, detoxification, neurological, and metabolism parameters (either enzymatic activity or gene expression/transcription) and changes on DNA in marine bivalves have been reported (e.g., Gonzalez-Rey and Bebianno 2014; Lacaze et al. 2015; Martin-Diaz et al. 2009).

Among the most commonly found pharmaceuticals in the environment is carbamazepine (CBZ), an antiepileptic drug used in the treatment of epilepsy, neuropathic pain, and psychiatric disorders (Ambrósio et al. 2002). CBZ has a global consumption volume of 1,014 tons per year (Zhang et al. 2008). As a consequence of low removal rates (<45%) in WWTPs (Bueno et al. 2012; Clara et al. 2004), it has been detected in the marine environment with concentrations between ng/L and few μg/L (Bayen et al. 2013; Biel-Maeso et al. 2018; Calisto et al. 2011; McEneff et al. 2014; Nödler et al. 2014). CBZ has also been reported in wild marine bivalves in concentrations up to 11 ng/g dry weight, revealing the potential for accumulation in these organisms (Klosterhaus et al. 2013; Moreno-González et al. 2016; Wille et al. 2011; Rodrigues et al. 2019). The present review summarizes the current knowledge of the biological effects exerted by CBZ in marine bivalves (either in early life stages or in adults) exposed to this drug under laboratory conditions (in vivo and in vitro studies). The effects of combined treatments with CBZ and other contaminants were not discussed in this review, and, thus, only effects derived from CBZ acting alone were here presented. The selection of marine bivalves was based on their relevant use as bioindicator species in toxicological assays worldwide. Major attention was given to the marine ecosystem, which is vulnerable to anthropogenic activities and much less investigated in comparison to the freshwater and terrestrial environments. Google Scholar, Scopus, ScienceDirect, and Web of Science were selected as databases for searching for literature.

2 Marine Bivalves Exposed to Carbamazepine Under Laboratory Conditions: Toxicological Effects

The toxicological effects of pharmaceutical drugs in the aquatic organisms exposed to laboratory conditions have been demonstrated in several works. However, major attention was devoted to the freshwater environment in contrast to the marine environment (Ebele et al. 2017; Gaw et al. 2014). Bivalves are exposed to pharmaceutical drugs discharged into the marine system near coastal sites where they naturally occur and where they can be harvested. Therefore, they provide valuable information not only on the exposure but also on the effects of drugs, contributing to the monitoring of the risk they represent to aquatic organisms and also to humans as shellfish consumers. The works available on the effects of CBZ in marine bivalves exposed to laboratory conditions are summarized in Table 1. Table 1 shows in vitro and in vivo studies that evaluated physiological, biochemical, and molecular endpoints in species of mussels, clams, and oysters to understand the effects of the drug.

Table 1 Biological effects of CBZ in marine bivalves exposed under laboratory conditions

Exposure concentration (μg/L)	Exposure (min, h, days) and depuration (D)	Species	Tissue/organ	Endpoints	Reference
0.3–9.0	96 h, 28 days	*M. galloprovincialis*	Whole tissue	Tissue drug concentration determination, physiological parameters, biochemical indicators of energy-related parameters and oxidative stress	Oliveira et al. (2017)
0.001, 0.01, 0.096	7 days	*P. viridis*	Hemolymph, gills, gonad, digestive gland	Biochemical indicators of immune functions, neurotoxicity, DNA damage, detoxification enzymes activity and of endocrine disruption	Juhel et al. (2017)
6.3	96 h	*M. galloprovincialis*	Hemolymph, digestive gland, gills	Biochemical indicators of oxidative stress, neurotoxicity, carbohydrate metabolism and metabolic enzymes, gene expression analysis	Brandts et al. (2018)
0.1, 10	7 days	*M. galloprovincialis*	Hemolymph, gills, digestive gland, mantle/gonad	Biochemical indicators of cell viability, accumulation of neutral lipids and lipofuscins, oxidative stress-related parameters, DNA damage, cyclic AMP content, PKA activity, mRNA expression of MXR-related genes	Martin-Diaz et al. (2009)
62,500–1,000,000	10 min	*M. galloprovincialis*	Gills, hemolymph	Biochemical indicators of neurotoxicity and biotransformation enzymes activity	Luis et al. (2016)
0.01, 10	48 h	*M. galloprovincialis*	Embryos	MXR efflux activity and ABCB/ABCC mRNA expressions	Franzellitti et al. (2017)
0.23–23,230	100 min	*M. galloprovincialis*	Hemolymph, hemocytes, gills	MXR efflux activity and ABCB/ABCC mRNA expressions, cytotoxicity effects	Franzellitti et al. (2016)
0.01–100	1 h	*M. galloprovincialis*	Hemocytes	Biochemical indicators of cell viability, oxidative stress-related parameters	Tsiaka et al. (2013)
1.5–75,000	21 h	*M. edulis*	Hemolymph	Biochemical indicators of cell viability, phagocytosis efficiency, DNA damage, ROS production	Lacaze et al. (2015)

0.03–9.0	96 h	*R. philippinarum* *R. decussatus*	Whole tissue	Tissue drug concentration determination, physiological parameters, biochemical indicators of oxidative stress and detoxification enzymes	Almeida et al. (2014)
0.03–9.0	28 days	*R. philippinarum*	Whole tissue	Tissue drug concentration determination, physiological parameters, biochemical indicators of oxidative stress, detoxification enzymes activity and energy-related parameters	Almeida et al. (2015)
1.0	28 days	*R. philippinarum*	Gills	Tissue drug concentration determination, biochemical indicators of oxidative stress and energy-related parameters, gene transcription	Almeida et al. (2018b)
30, 50	14 days	*R. decussatus*	Gills	Tissue drug and metabolites concentration determination, biochemical indicators of oxidative stress	Abdelhafidh et al. (2018)
0.1–50	35 days	*R. philippinarum*	Hemolymph	Biochemical indicator of cell viability (lysosomal membrane stability)	Aguirre-Martínez et al. (2013)
1	28 days	*R. philippinarum*	Whole tissue	Tissue drug concentration determination, biochemical indicators of oxidative stress, biotransformation enzymes activity and energy-related parameters	Almeida et al. (2018a)
0.1–50	14 days	*R. philippinarum*	Hemolymph, digestive gland	Biochemical indicator of cell viability, biotransformation enzymes activity, oxidative stress, neurotoxicity, DNA damage	Aguirre-Martínez et al. (2016)
15	14 days + 7 D	*R. philippinarum*	Gills, digestive gland	Physiological parameters, biochemical indicators of oxidative stress and neurotoxicity, biotransformation enzymes activity	Trombini et al. (2019)
0.03–9.0	96 h	*S. plana*	Whole tissue	Tissue drug concentration determination, biochemical indicators of oxidative stress, biotransformation enzymes activity and energy-related parameters	Almeida et al. (2017a)

(continued)

Table 1 (continued)

Exposure concentration (μg/L)	Exposure (min, h, days) and depuration (D)	Species	Tissue/organ	Endpoints	Reference
4, 8	14 days + 14 D	*S. plana*	Whole tissue	Tissue drug concentration determination, biochemical indicators of oxidative stress, biotransformation enzymes activity and energy-related parameters	Almeida et al. (2017c)
3	28 days	*S. plana*	Whole tissue	Tissue drug concentration determination, elements (Na, K) quantification, biochemical indicators of oxidative stress, biotransformation enzymes activity and energy-related parameters	Freitas et al. (2016)
3	96 h	*S. plana*	Whole tissue	Tissue drug concentration determination, physiological parameters, elements (Na, K) quantification, biochemical indicators of oxidative stress, biotransformation enzymes activity and energy-related parameters	Freitas et al. (2015a)
0.3–9.0	28 days	*S. plana*	Whole tissue	Tissue drug concentration determination, biochemical indicators of oxidative stress, biotransformation enzymes activity and energy-related parameters	Freitas et al. (2015b)
10,000–50,000	36 h (embryotoxicity test) 24 h (metamorphosis test)	*C. gigas*	Embryos, pediveliger larvae	Embryotoxicity test, metamorphosis test	Di Poi et al. (2018)

2.1 Mussels

2.1.1 CBZ Effects on Physiological Processes

The application of physiological parameters has been performed to study the effects of CBZ on exposed marine mussels. Oliveira et al. (2017) showed a reduction in the condition (CI) and gonadosomatic index (GSI) in mussels (*Mytilus galloprovincilais*) acutely (96 h) and chronically (28 days) exposed to different CBZ concentrations (0.3–9.0 μg/L). These authors suggested that, to cope with the exposure to CBZ, mussels maintained their valves closed to avoid the accumulation of the drug. Thus, the decrease of the feeding capacity will be expected, with a consequent reduction of CI values. With regard to the GSI values, Oliveira et al. (2017) suggested that the reduction observed at CBZ 9.0 μg/L could be a consequence of a slowdown in the mussels' metabolic capacity, resulting in a decrease of the expenditure of energy reserves, not being used in the somatic growth and reproduction nor in defense and repair mechanisms. However, it is important to refer that the observed changes in the physiological parameters (CI and GSI) were only observed at higher CBZ concentrations (6.0 and 9.0 μg/L) that are not common in the marine environment.

2.1.2 CBZ Effects on the Biotransformation and Detoxification Processes

Carbamazepine, like many pharmaceutical drugs, requires biotransformation for elimination. The cytochrome P450 system (CYP 450), namely, the P450 3A4 (CYP3A4), is the most important family of enzymes involved in drug biotransformation (Anzenbacher and Anzenbacherova 2001). In humans, CBZ is predominantly metabolized by CYP3A4 to 10,11-dihydro-10,11-epoxycarbamazepine (CBZ-epoxide), which is pharmacologically active. CBZ-epoxide is eliminated by hydroxylation of the aromatic ring to 10,11-trans-dihydro-10,11-dihydroxycarbamazepine (CBZ diol, the main metabolite excreted in urine) or by conjugation to form O-glucuronides (Bueno et al. 2013; Houeto et al. 2012). Besides CYP 450 enzymes (phase I of biotransformation), glutathione-S-transferases (GSTs) are another family of biotransformation enzymes (phase II of biotransformation). However, GSTs also possess a role in the defense against oxidative damage and peroxidative products of DNA and lipids and are commonly studied for effects related with oxidative stress (Regoli and Giuliani 2014). The activity of the referred phase I and II biotransformation enzymes has been studied in mussels exposed to CBZ. For example, Juhel et al. (2017) showed an increase of 7-ethoxyresorufin-O-deethylase activity in *Perna viridis* mussels at CBZ concentrations of 96 ng/L (7 days exposure). These results were related with the increase of biosynthesis of a CYP1A isoenzyme, an enzyme from the CYP 450 family. Also, the upregulation of CYP11 gene expression in *M. galloprovincialis* exposed to CBZ (6.3 μg/L, 96 h exposure period) demonstrated the involvement of CYP 450 system enzymes in the detoxification of this drug in mussels (Brandts et al. 2018). However, conversely to the previous studies, Oliveira et al. (2017) found no induction in the activity of CYP3A4 in *M. galloprovincialis* exposed to CBZ (0.3–9 μg/L) for 96 h and 28 days. Even though different exposure conditions were applied in the abovementioned studies, results obtained seemed to indicate that CBZ suffers detoxification in mussels, possibly being transformed in metabolites. These findings are in accordance with the study of Boillot et al. (2015) that showed the occurrence of CBZ metabolites (carbamazepine-10,11-epoxide and acridine) in *M. galloprovincialis* exposed to 10 μg/L for 1 week. With regard to phase II biotransformation enzymes (GSTs), Martin-Diaz et al. (2009) found an induction of the enzymatic activity in digestive gland and mantle/gonads of *M. galloprovincialis* exposed for 7 days to CBZ (0.1 and 10 μg/L). Also, Luis et al. (2016), applying an in vitro assay (CBZ 62.5–100 mg/L, 10 min of incubation) with *M. galloprovincialis*, showed only changes in the activity of GSTs (a decrease) for non-environmental CBZ concentrations, higher than 250 mg/L.

In order to evaluate the detoxification/excretion processes in mussels' early life stages and adults, Franzellitti et al. (2016, 2017) studied the regulation of ATP-binding cassette (ABC) transport proteins, an integral part of the multixenobiotic resistance (MXR) system, in aquatic organisms. These transporters are involved in the detoxification of harmful compounds by marine bivalves, preventing the accumulation and toxic effects. In a 48 h embryotoxicity test with early life stages of *M. galloprovincialis* exposed to CBZ (0.01 and 10 μg/L), Franzellitti et al. (2017) assessed the activity (for P-glycoprotein and MXR-related proteins) and mRNA expression (for the ABC transporters: ABCB and ABCC) of the MXR efflux transporters. CBZ interacted with the MXR transporters in mussel embryos, with effects changing with the progression of embryo development. Indeed, CBZ effects on MXR activity were observed in trocophores (10 μg/L) and D-veligers (0.01 and 10 μg/L), being D-veligers referred as the most sensitive stage to CBZ-induced MXR modulation. CBZ was mentioned as a possible substrate for P-glycoprotein and as a negative transcription regulator for MXR-related proteins. In another study, Franzellitti et al. (2016) studied the activity and transcription of MXR transporters in the hemocytes of adult *M. galloprovincialis* exposed to CBZ (0.23 μg/L–23.23 mg/L) concentrations. The positive regulation of ABCB transcription in mussels exposed to CBZ was related with the increased requirements for the synthesis of novel P-glycoprotein transporters to extrude CBZ from cells. In general, the previous studies suggest that mussels are capable to metabolize and excrete CBZ at some extent, a fact that should be taken into consideration, for example, when analyzing wild mussels.

2.1.3 CBZ Effects on Oxidative Stress-Related Parameters

In addition to the investigation of physiological and biotransformation/detoxification processes, oxidative stress-related analyses have been commonly applied in invertebrates exposed to xenobiotics, including drugs. Indeed, ROS (reactive oxygen species) can lead to cellular damage, DNA strand break formation, and inactivation of enzymes (Regoli and Giuliani 2014). Also, ROS impacts have been related with the development of neurotoxicity in vertebrates due to the high rate of oxygen utilization by the central nervous system, poor concentration of antioxidants, and high content of polyunsaturated lipids (Sayre et al. 2008), a response that may also occur in invertebrate organisms. Several authors, through biochemical analyses, denoted the induction of oxidative stress promoted by CBZ due to the generation of ROS as a result of biotransformation and metabolic processes. In particular, the development of an oxidative stress response was demonstrated in marine mussels exposed to CBZ. Tsiaka et al. (2013) reported an increase in cell death and in the levels of superoxide anions and nitric oxides in *M. galloprovincialis* hemocytes exposed to CBZ (0.01–100 μg/L) for 1 h. The authors referred that the accumulation of CBZ within lysosomes was responsible for lysosomal membrane impairment and the leak of hydrolytic enzymes into the cytosol, inducing oxidative damage. The development of oxidative stress, although without exerting oxidative damage, was also observed in the previously mentioned study by Oliveira et al. (2017). These authors showed differences mainly in the antioxidant defense enzymatic activity (superoxide dismutase and catalase), with a decrease in activity in mussels exposed to CBZ by comparison with the control. These results were related with an insufficient production of ROS exerted by the drug to induce these defenses. However, comparing sampling periods, a higher activity of these enzymes was observed in the acute test in comparison with the chronic exposure. Moreover, different trends were observed in the metabolic capacity of exposed mussels comparing the two experimental periods: increase in the electron transport system activity (ETS) with the increasing concentrations in the acute exposure and an opposite response in the chronic exposure. The decrease in the metabolic activity (reduced ETS), possibly due to a reduction in filter feeding activity, and even with some associated costs for the organisms (decrease in GSI and CI), was possibly enough to avoid the damage of CBZ. In general, Oliveira et al. (2017) suggested that mussels exposed to CBZ under a chronic exposure, in comparison with acutely exposed mussels, appeared to overcome the stress induced by the drug. Other studies found, however, that CBZ induced oxidative stress in *M. galloprovincialis* exposed to CBZ 0.1 and 10 μg/L for 7 days through a reduction in the integrity of the lysosome membrane, an increase in the lipid peroxidation (LPO) levels, and an increase in the activity of an antioxidant enzyme (catalase) (Martin-Diaz et al. 2009). On the other hand, Luis et al. (2016) found that CBZ inhibited the activity of acetylcholinesterase (related with neurotoxicity) in *M. galloprovincialis* cells, although at much higher concentrations (250–1,000 mg/L), compared with the previous studies. However, in the study performed by Juhel et al. (2017), at a low concentration of 96 ng/L of CBZ, a strong inhibition of acetylcholinesterase was found in *P. viridis* mussels exposed for 7 days. The CBZ-induced neurotoxicity was related with the alteration of physiological functions in the organisms (e.g., feeding behavior, growth and survival).

2.1.4 CBZ Effects on Multiple General Stress Responses

In addition to oxidative stress, CBZ affected other biological processes in marine mussels. For example, Lacaze et al. (2015) observed that CBZ induced significant genotoxic aberrations in the hemocytes of *M. edulis* exposed to a concentration range between 0.0015 and 75 mg/L for 21 h. The changes in the transcription of genes related with DNA repair, cell stress response, and innate immunity were also demonstrated by the previous authors (Juhel et al. 2017). For example, regarding DNA damage repair genes, the expression of gaadd45α was increased in mussels exposed to CBZ. Also, an upregulation of the gene expression of cathepsins was found in mussels exposed to CBZ. The upregulation of cathepsins was related with an involvement in lysosomal digestion and detoxification, possibly reflecting the lysosomal destabilization caused by ROS, as already referred in the previous studies.

2.1.5 CBZ Effects on the Mode of Action-Related Processes and Metabolic Pathways

Concerned with whether CBZ interacts in mussels with the same cellular targets as its therapeutic action in humans, some authors investigated the mechanisms of action of the drug in mussels. Although the main target of CBZ in humans are the voltage-dependent Na^+ channels, this drug has also been referred to interact with Ca^+ and K^+ channels (Ambrósio et al. 2002); to modulate the release, uptake, and receptor binding of neurotransmitters as GABA (gamma-aminobutyric acid) (Ambrósio et al. 2002; Motohashi et al. 1989); to inhibit adenylate cyclase activity (Chen et al. 1996); to decrease arachidonic acid turnover (Bazinet et al. 2006); and to inhibit the histone deacetylases (Beutler et al. 2005). Taking this into account, the study of Martin-Diaz et al. (2009) measured the activity of adenylate cyclase and concomitant levels of intracellular cAMP in *M. galloprovincialis* after CBZ exposure. The authors observed a significant reduction of cAMP content and the adenylate cyclase activity in mussels' tissues at 0.1 and 10 μg/L, revealing that CBZ interacts with a specific molecular target that is also conserved in mussels. This information is particularly important since cAMP is related with several functions in bivalves, such as the regulation of glycogen breakdown and concomitant control of gonad maturation (Fabbri and Capuzzo 2010).

2.2 *Clams*

2.2.1 CBZ Effects on Physiological Processes

The impact of CBZ on the physiological processes of clams was, for the first time, shown by Almeida et al. (2014). These authors reported the potential of clams *Ruditapes philippinarum* and *R. decussatus* to decrease the filtration rate, as observed by the decrease in the clearance rate of organisms exposed to CBZ (0–9.0 μg/L) for 96 h, thus affecting the organisms' metabolic processes. A chronic study (28 days) by the same authors (Almeida et al. 2015) confirmed the same trend in *R. philippinarum* exposed to the same concentration range. However, in both studies, the major effects on the clearance rate were observed at the highest CBZ concentration (9.0 μg/L), while at lower concentrations, the clearance rate was, in general, similar to non-exposed clams. These results are in agreement with the previously referred study of Oliveira et al. (2017), performed with mussels. These authors, although not testing the effects of CBZ on the feeding rate, suggested that, to cope with the exposure to CBZ, possibly mussels decreased their feeding rate by maintaining their valves closed. It is still important to refer that the effects exerted by CBZ on the physiological parameters of both clams and mussels occurred mainly at 9.0 μg/L, a concentration that is not known in the marine environment.

2.2.2 CBZ Effects on the Biotransformation and Detoxification Processes

As for mussels, the investigations of phase I and phase II biotransformation enzymes have been performed through biochemical and molecular endpoints. Almeida et al. (2018b) found a decrease in CYP3 mRNA transcription in the gills of CBZ-exposed *R. philippinarum* (1.0 μg/L, 28 days). However, these authors expected an upregulation of CYP3 transcription since CBZ is known to be metabolized after consumption in humans, with the involvement of CYP 450 system (namely, CYP3A4 and CYP2C8 enzymes) (Magnusson et al. 2008). At a wide concentration range of CBZ (0.1–50 μg/L, 14 days exposure), Aguirre-Martínez et al. (2016) showed the increase of ethoxyresorufin-O-deethylase (at all exposure concentrations) and dibenzylfluorescein dealkylase (only at 50 μg/L) activities in *R. philippinarum*, which are CYP enzymes corresponding to CYP1A and CYP3A isoforms, respectively. Also, some studies endeavored to address the activity of phase I and II biotransformation enzymes (CYP 450, GSTs) in CBZ-exposed clams, aiming to investigate the mechanisms of detoxification in these organisms. The previously referred Almeida et al. (2014) study showed that an acute exposure to CBZ caused significant changes in the activity of CYP3A4 and GSTs in *R. philippinarum* and *R. decussatus*, but no clear concentration-dependent trends were observed. In the long-term assay of Aguirre-Martínez et al. (2016), GSTs activities were only significantly increased at the highest exposure concentration (50 μg/L). Also, in *Scrobicularia plana*, Freitas et al. (2015a) found distinct results in CBZ-exposed (3 μg/L, 96 h) clams collected from a contaminated and pristine area of a marine system. The authors found an increase in the activity of GSTs in CBZ-exposed clams from the contaminated area, while in the CBZ-exposed clams from the pristine area, no significant changes on GSTs activity were found. However, as GSTs are also involved with the inactivation of endogenous secondary metabolites formed during oxidative stress, Freitas et al. (2015a) suggested a defense mechanism from the clams of the contaminated area to degrade the higher peroxidized lipids and not being related with the detoxification processes. The GSTs involvement in the oxidative stress will be referred in the next topic. Other studies found no clear concentration-dependent effects of CBZ in the activity of GSTs in *S. plana* acutely and chronically exposed to a wide range of concentrations of the drug (0.03–9.0 μg/L) (Almeida et al. 2017a, b; Freitas et al. 2015b, 2016). Despite the no clear effects of CBZ on phase I and phase II biotransformation enzymes in the previously referred studies, Abdelhafidh et al. (2018) showed the occurrence of one CBZ metabolite (3-hydroxy-CBZ) in *R. decussatus* clams exposed to CBZ non-environmental concentrations (30 and 50 μg/L) for 14 days, indicating the capability of clams to metabolize and excrete CBZ at higher concentrations.

2.2.3 CBZ Effects on Oxidative Stress-Related Parameters

The effects of CBZ in the oxidative processes of marine clams were also reported for a few species (*R. philippinarum*, *R. decussatus*, and *S. plana*). As in mussels, CBZ (0.1–50 μg/L) also induced a significant reduction in lysosomal membrane stability of *R. philippinarum* clams exposed for 35 days (Aguirre-Martínez et al. 2013). The previous results were related with a disruption of the lysosome membrane with the leakage of the lysosome contents into the cytosol and consequent damage to cells, a response already reported in mussels (Tsiaka et al. 2013). Thus, authors investigated the occurrence of oxidative stress in clams exposed to CBZ. For that, indicators of cellular damage as lipid peroxidation levels and activities of enzymes related with antioxidant/biotransformation systems and neurotoxicity were evaluated. Almeida et al. (2014) showed that CBZ induced oxidative stress (increased antioxidant defense and biotransformation enzymatic activity and LPO) in *R. philippinarum* and *R. decussatus* clams. The authors referred that these two species showed different responses to CBZ, in which *R. decussatus* by opposition to *R. philippinarum* was able to induce the antioxidant system and prevent the oxidative damage of CBZ. Also, in the study by Abdelhafidh et al. (2018), *R. decussatus* clams exposed to CBZ non-environmental concentrations showed an oxidative stress (increased activity of antioxidant and biotransformation enzymes and LPO levels) response. Although different exposure concentrations were applied in the abovementioned studies (41-fold difference comparing the highest exposure concentration), results seems to indicate that *R. decussatus* and *R. philippinarum* possess different stress limits. However, in the presence of higher CBZ concentrations, both species are negatively affected. Under long-term exposures, authors also found the induction of oxidative stress. The previously referred study of Aguirre-Martínez et al. (2016) also showed that CBZ induced oxidative stress (e.g., increased LPO levels) and neurotoxicity (decreased acetylcholinesterase activity) at all exposure concentrations tested (0.1–50 μg/L) in *R. philippinarum* exposed for 14 days. Trombini et al. (2019) evaluated biochemical responses related with oxidative stress in *R. philippinarum* exposed to CBZ (15 μg/L) in several days along a 14-day exposure duration followed by a 7-day depuration period. The authors showed that CBZ induced oxidative changes particularly after short-term exposure, during the first week of exposure and particularly during the first 24 h. However, by the end of exposure and especially after depuration, the authors showed that clams were able to cope with these imbalances by recovering their general oxidative status. Almeida et al. (2015) also found the induction of oxidative stress in clams *R. philippinarum* exposed to CBZ (0–9.0 μg/L) for 28 days, as revealed by the increased LPO levels (only at 3.0 μg/L) and antioxidant defense enzyme activity (superoxide dismutase). According to the authors, at higher concentrations, CBZ was somehow signalized as harmful. To limit the exposure to CBZ contamination, clams reduced their metabolic capacity and drug uptake as observed by the reduction in the clearance rate and the electron transport system activity. Despite this strategy, the compromised activity of antioxidant (superoxide dismutase) and biotransformation (GSTs) enzymes resulted in cellular damage. At lower concentrations, despite the clams' increased filtration and metabolic activity, the antioxidant and biotransformation enzymes were also promptly activated, avoiding oxidative damage. However, oxidative damage was also observed in *R. philippinarum* exposed to an intermediate CBZ exposure concentration (1.0 μg/L) for a similar period of time (28 days) (Almeida et al. 2018a). In addition to *R. philippinarum* and *R. decussatus* clams, *S. plana* clams were also tested for possible negative effects of CBZ when subjected to this contamination under laboratory conditions. CBZ was shown to exert mainly oxidative damage in *S. plana* in several studies covering acute and chronic exposures (96 h and 28 days) and a wide range of CBZ concentrations (0.03–9.0 μg/L) (Almeida et al. 2017a, b, c; Freitas et al. 2015a, b, 2016).

2.2.4 CBZ Effects on the Mode of Action-Related Processes

As for mussels, the investigation of the CBZ effects related with its mode of action in humans was also performed for clams, but only in one study. The exposure of *R. philippinarum* for 28 days to CBZ 1.0 μg/L was shown to change the transcription of genes related to the mode of action of the drug (e.g., metabolism – GABA receptor) (Almeida et al. 2018b). The upregulation of GABA receptor mRNA transcription of clams exposed to CBZ was related with the capability of CBZ to modulate the release, uptake, and receptor binding of neurotransmitters like GABA in vertebrates (Motohashi et al. 1989). Despite the changes on GABA receptor mRNA transcription, Almeida et al. (2018b) found no oxidative stress induction in the clams exposed to 1.0 μg/L after 28 days.

2.3 *Oysters*

The effects of CBZ (10–50 mg/L) on oysters were demonstrated by Di Poi et al. (2018) through an embryotoxicity and metamorphosis test on *Crassostrea gigas*. Dose-dependent effects on the effective concentration (EC) values were observed; however, for each tested concentration (10, 20, and 50 mg/L), lower EC values were found on the 36 h embryotoxicity test (5.27–17.48 mg/L) compared with the 24 h metamorphosis test (10.78–22.83 mg/L). The authors related the toxic effect on the larval level with possible impairments on the population, with economic consequences. Despite this being the only study where the effects of CBZ were evaluated in oyster species, it is possible to observe that the EC values are in the mg/L range, a concentration range that was never found in marine systems (in general up to 1 μg/L). Also, although different endpoints were assessed by Franzellitti et al. (2017), it is possible to infer that mussels' early stages are possibly more sensitive to CBZ than oysters' early stages.

Considering all the above studies on the biological effects of CBZ in marine bivalves, overall, CBZ affected a few species of these organisms (mussels, *M. galloprovincialis*, *M. edulis*, and *P. viridis*; clams, *R. philippinarum*, *R. decussatus*, and *S. plana*; oysters, *C. gigas*) using acute and chronic tests (in vivo and in vitro studies), performed with environmental and non-environmental concentrations. Biological effects related with the individual and sub-individual level were applied in a non-targeted and targeted approach (related with the mode of action of the drug). Overall, in marine bivalves, CBZ changed physiological responses (filter feeding activity, condition index, gonadosomatic index) mainly at non-environmentally relevant concentrations and induced cytotoxicity, oxidative stress and change of related parameters, immunotoxicity, genotoxicity and interacted with targets related with the specific mode of action. However, despite the application of different exposure conditions (concentrations, cell types, biological endpoints) and methodological approaches which make difficult a better understanding of the impacts of the drug, in general, studies reported the occurrence of oxidative stress in marine bivalves as the main effect exerted by CBZ, even at marine environmentally relevant concentrations.

3 Conclusions and Future Perspectives

Carbamazepine (CBZ), an antiepileptic drug, has been detected in several aquatic systems with proven negative effects for inhabiting organisms. The effects resulting from laboratory exposures to CBZ were reviewed here in order to understand possible impacts derived from the exposure to this drug under environmental conditions. Bivalves were selected to address this topic as they represent key organisms in aquatic systems.

The investigation of the effects of CBZ in marine bivalves (either adults or early life stages) has been assessed acutely and chronically at the individual and sub-individual levels. In general, sub-lethal effects related to cytotoxicity, oxidative stress and related parameters, immunotoxicity, and genotoxicity were the most investigated. Overall, concerning the effects of CBZ exposure in marine bivalves, in general, studies reported the occurrence of oxidative stress, as the main impact. The exposure concentrations ranged between 0.03 μg/L and 50 mg/L (in vivo and in vitro studies), some of them being similar to the CBZ concentrations found in the marine environment (up to ~1 μg/L). Although some of the concentrations tested under laboratory conditions were higher than the ones found in the marine environment, some of them cover the range of concentrations found in freshwater systems. Also, the exposure time of the organisms to CBZ under laboratory conditions ranged between 96 h and 28 days (for adult bivalves exposed in vivo), while for the use of specific cell types (e.g., hemocytes), the exposure periods in the in vitro assays were shorter. In future studies, an integrative approach of the chronic effects at both individual and sub-individual levels, tested at environmental concentrations/scenarios, should be included, as well as the determination of the concentrations and drug metabolites in the animals exposed to those scenarios.

Acknowledgments Ângela Almeida benefited from PhD grant (SFRH/BD/110218/2015) given by the National Funds through the Portuguese Science Foundation (FCT), supported by FSE and Programa Operacional Capital Humano (POCH) and European Union. Rosa Freitas was funded by national funds (OE), through FCT-Fundação para a Ciência e a Tecnologia, I.P., in the scope of the framework contract foreseen in the numbers 4, 5, and 6 of the article 23, of the Decree-Law 57/2016, of August 29, changed by Law 57/2017, of July 19. This work was also financially supported by the project BISPECIAl: BIvalveS under Polluted Environment and ClImate chAnge PTDC/CTA-AMB/28425/2017 (POCI-01-0145-FEDER-028425) funded by FEDER, through COMPETE2020 – Programa Operacional Competitividade e Internacionalização (POCI), and by national funds (OE), through FCT/MCTES. Thanks are also due to Integrated Programme of SR&TD "Smart Valorization of Endogenous Marine Biological Resources Under a Changing Climate" (reference Centro-01-0145-FEDER-000018), co-funded by Centro 2020 program, Portugal 2020, European Union, through the European Regional Development Fund. Thanks are also due for the financial support to CESAM (UIDB/50017/2020 + UIDP/50017/2020). This manuscript was English revised by Gabinete de Tradução, Revisão e Comunicação Técnica, University of Aveiro.

Conflict of Interest The authors whose names are listed immediately below certify that they have NO affiliations with or involvement in any organization or entity with any financial interest (such as honoraria; educational grants; participation in speakers' bureaus; membership, employment, consultancies, stock ownership, or other equity interest; and expert testimony or patent-licensing arrangements) or non-financial interest (such as personal or professional relationships, affiliations, knowledge, or beliefs) in the subject matter or materials discussed in this manuscript.

References

Abdelhafidh K, Ali M, Hassen K, Badreddine S, Jaume A, Sandra P, Ethel E, Damià B, Hamouda B, Ezzeddine M (2018) Uptake and metabolism of carbamazepine (CBZ) by clam *Ruditapes decussatus* and its effects in biochemical responses. Xenobiotica 48:727–733

Aguirre-Martínez GV, Buratti S, Fabbri E, DelValls AT, Martín-Díaz ML (2013) Using lysosomal membrane stability of haemocytes in *Ruditapes philippinarum* as a biomarker of cellular stress to assess contamination by caffeine, ibuprofen, carbamazepine and novobiocin. J Environ Sci 25:1408–1418

Aguirre-Martínez GV, DelValls TA, Martín-Díaz ML (2016) General stress, detoxification pathways, neurotoxicity and genotoxicity evaluated in *Ruditapes philippinarum* exposed to human pharmaceuticals. Ecotoxicol Environ Saf 124:18–31

Almeida Â, Calisto V, Esteves VI, Schneider RJ, Soares AM, Figueira E, Freitas R (2014) Presence of the pharmaceutical drug carbamazepine in coastal systems: effects on bivalves. Aquat Toxicol 156:74–87

Almeida Â, Freitas R, Calisto V, Esteves VI, Schneider RJ, Soares AM, Figueira E (2015) Chronic toxicity of the antiepileptic carbamazepine on the *clam Ruditapes philippinarum*. Comp Biochem Physiol C Toxicol Pharmacol 172:26–35

Almeida Â, Calisto V, Domingues MRM, Esteves VI, Schneider RJ, Soares AM, Figueira E, Freitas R (2017a) Comparison of the toxicological impacts of carbamazepine and a mixture of its photodegradation products in *Scrobicularia plana*. J Hazard Mater 323:220–232

Almeida Â, Calisto V, Esteves VI, Schneider RJ, Soares AM, Figueira E, Freitas R (2017b) Ecotoxicity of the antihistaminic drug cetirizine to *Ruditapes philippinarum* clams. Sci Total Environ 601:793–801

Almeida Â, Calisto V, Esteves VI, Schneider RJ, Soares AM, Figueira E, Freitas R (2017c) Toxicity associated to uptake and depuration of carbamazepine in the clam *Scrobicularia plana* under a chronic exposure. Sci Total Environ 580:1129–1145

Almeida Â, Calisto V, Esteves VI, Schneider RJ, Soares AM, Figueira E, Freitas R (2018a) Effects of single and combined exposure of pharmaceutical drugs (carbamazepine and cetirizine) and a metal (cadmium) on the biochemical responses of R. *philippinarum*. Aquat Toxicol 198:10–19

Almeida Â, Freitas R, Calisto V, Esteves VI, Schneider RJ, Soares AM, Figueira E, Campos B, Barata C (2018b) Effects of carbamazepine and cetirizine under an ocean acidification scenario on the biochemical and transcriptome responses of the clam *Ruditapes philippinarum*. Environ Pollut 235:857–868

Ambrósio AF, Soares-da-Silva P, Carvalho CM, Carvalho AP (2002) Mechanisms of action of carbamazepine and its derivatives, oxcarbazepine, BIA 2-093, and BIA 2-024. Neurochem Res 27:121–130

Anzenbacher P, Anzenbacherova E (2001) Cytochromes P450 and metabolism of xenobiotics. Cell Mol Life Sci 58:737–747

Bayen S, Zhang H, Desai MM, Ooi SK, Kelly BC (2013) Occurrence and distribution of pharmaceutically active and endocrine disrupting compounds in Singapore's marine environment: influence of hydrodynamics and physical–chemical properties. Environ Pollut 182:1–8

Bazinet RP, Rao JS, Chang L, Rapoport SI, Lee H-J (2006) Chronic carbamazepine decreases the incorporation rate and turnover of arachidonic acid but not docosahexaenoic acid in brain phospholipids of the unanesthetized rat: relevance to bipolar disorder. Biol Psychiatry 59:401–407

Beutler AS, Li S, Nicol R, Walsh MJ (2005) Carbamazepine is an inhibitor of histone deacetylases. Life Sci 76:3107–3115

Biel-Maeso M, Baena-Nogueras RM, Corada-Fernández C, Lara-Martín PA (2018) Occurrence, distribution and environmental risk of pharmaceutically active compounds (PhACs) in coastal and ocean waters from the Gulf of Cadiz (SW Spain). Sci Total Environ 612:649–659

Boillot C, Bueno MM, Munaron D, Le Dreau M, Mathieu O, David A, Fenet H, Casellas C, Gomez E (2015) In vivo exposure of marine mussels to carbamazepine and 10-hydroxy-10, 11-dihydro-carbamazepine: bioconcentration and metabolization. Sci Total Environ 532:564–570

Brandts I, Teles M, Gonçalves AP, Barreto A, Franco-Martinez L, Tvarijonaviciute A, Martins MA, Soares A, Tort L, Oliveira M (2018) Effects of nanoplastics on *Mytilus galloprovincialis* after individual and combined exposure with carbamazepine. Sci Total Environ 643:775–784

Brausch JM, Connors KA, Brooks BW, Rand GM (2012) Human pharmaceuticals in the aquatic environment: a review of recent toxicological studies and considerations for toxicity testing. In: Whitacre DM (ed) Reviews of environmental contamination and toxicology, Reviews of environmental contamination and toxicology, vol 218. Springer, Boston, pp 1–99

Bueno MM, Gomez MJ, Herrera S, Hernando MD, Agüera A, Fernández-Alba AR (2012) Occurrence and persistence of organic emerging contaminants and priority pollutants in five sewage treatment plants of Spain: two years pilot survey monitoring. Environ Pollut 164:267–273

Bueno MM, Boillot C, Fenet H, Chiron S, Casellas C, Gómez E (2013) Fast and easy extraction combined with high resolution-mass spectrometry for residue analysis of two anticonvulsants and their transformation products in marine mussels. J Chromatogr A 1305:27–34

Calisto V, Bahlmann A, Schneider RJ, Esteves VI (2011) Application of an ELISA to the quantification of carbamazepine in ground, surface and wastewaters and validation with LC–MS/MS. Chemosphere 84:1708–1715

Chen G, Pan B, Hawver DB, Wright CB, Potter WZ, Manji HK (1996) Attenuation of cyclic AMP production by carbamazepine. J Neurochem 67:2079–2086

Clara M, Strenn B, Kreuzinger N (2004) Carbamazepine as a possible anthropogenic marker in the aquatic environment: investigations on the behaviour of carbamazepine in wastewater treatment and during groundwater infiltration. Water Res 38:947–954

Di Poi C, Costil K, Bouchart V, Halm-Lemeille M-P (2018) Toxicity assessment of five emerging pollutants, alone and in binary or ternary mixtures, towards three aquatic organisms. Environ Sci Pollut Res 25:6122–6134

Ebele AJ, Abdallah MA-E, Harrad S (2017) Pharmaceuticals and personal care products (PPCPs) in the freshwater aquatic environment. Emerg Contam 3:1–16

Fabbri E, Capuzzo A (2010) Cyclic AMP signaling in bivalve molluscs: an overview. J Exp Zool A Ecol Genet Physiol 313:179–200

Fabbri E, Franzellitti S (2016) Human pharmaceuticals in the marine environment: focus on exposure and biological effects in animal species. Environ Toxicol Chem 35:799–812

Franzellitti S, Striano T, Valbonesi P, Fabbri E (2016) Insights into the regulation of the MXR response in haemocytes of the Mediterranean mussel (*Mytilus galloprovincialis*). Fish Shellfish Immunol 58:349–358

Franzellitti S, Striano T, Pretolani F, Fabbri E (2017) Investigating appearance and regulation of the MXR phenotype in early embryo stages of the Mediterranean mussel (*Mytilus galloprovincialis*). Comp Biochem Physiol C Toxicol Pharmacol 199:1–10

Freitas R, Almeida Â, Calisto V, Velez C, Moreira A, Schneider RJ, Esteves VI, Wrona FJ, Soares AM, Figueira E (2015a) How life history influences the responses of the clam *Scrobicularia plana* to the combined impacts of carbamazepine and pH decrease. Environ Pollut 202:205–214

Freitas R, Almeida Â, Pires A, Velez C, Calisto V, Schneider RJ, Esteves VI, Wrona FJ, Figueira E, Soares AM (2015b) The effects of carbamazepine on macroinvertebrate species: comparing bivalves and polychaetes biochemical responses. Water Res 85:137–147

Freitas R, Almeida Â, Calisto V, Velez C, Moreira A, Schneider RJ, Esteves VI, Wrona FJ, Figueira E, Soares AMVM (2016) The impacts of pharmaceutical drugs under ocean acidification: new data on single and combined long-term effects of carbamazepine on *Scrobicularia plana*. Sci Total Environ 541:977–985

Gaw S, Thomas KV, Hutchinson TH (2014) Sources, impacts and trends of pharmaceuticals in the marine and coastal environment. Philos Trans R Soc B Biol Sci 369:20130572

Gonzalez-Rey M, Bebianno MJ (2014) Effects of non-steroidal anti-inflammatory drug (NSAID) diclofenac exposure in mussel *Mytilus galloprovincialis*. Aquat Toxicol 148:221–230

Halling-Sørensen B, Nielsen SN, Lanzky PF, Ingerslev F, Lützhøft HH, Jørgensen S (1998) Occurrence, fate and effects of pharmaceutical substances in the environment–a review. Chemosphere 36:357–393

Honkoop PJ, Luttikhuizen PC, Piersma T (1999) Experimentally extending the spawning season of a marine bivalve using temperature change and fluoxetine as synergistic triggers. Mar Ecol Prog Ser 180:297–300

Houeto P, Carton A, Guerbet M, Mauclaire A-C, Gatignol C, Lechat P, Masset D (2012) Assessment of the health risks related to the presence of drug residues in water for human consumption: application to carbamazepine. Regul Toxicol Pharmacol 62:41–48

Juhel G, Benotti S, Goh C, Lee WK, Kelly BC (2017) Use of a suite of biomarkers to assess the effects of carbamazepine, bisphenol A, atrazine, and their mixtures on green mussels, Perna viridis. Environ Toxicol Chem 36:429–441

Klosterhaus SL, Grace R, Hamilton MC, Yee D (2013) Method validation and reconnaissance of pharmaceuticals, personal care products, and alkylphenols in surface waters, sediments, and mussels in an urban estuary. Environ Int 54:92–99

Lacaze E, Pédelucq J, Fortier M, Brousseau P, Auffret M, Budzinski H, Fournier M (2015) Genotoxic and immunotoxic potential effects of selected psychotropic drugs and antibiotics on blue mussel (*Mytilus edulis*) hemocytes. Environ Pollut 202:177–186

Luis LG, Barreto Â, Trindade T, Soares AM, Oliveira M (2016) Effects of emerging contaminants on neurotransmission and biotransformation in marine organisms—an in vitro approach. Mar Pollut Bull 106:236–244

Magnusson MO, Dahl M-L, Cederberg J, Karlsson MO, Sandström R (2008) Pharmacodynamics of carbamazepine-mediated induction of CYP3A4, CYP1A2, and Pgp as assessed by probe substrates midazolam, caffeine, and digoxin. Clin Pharmacol Ther 84:52–62

Martin-Diaz L, Franzellitti S, Buratti S, Valbonesi P, Capuzzo A, Fabbri E (2009) Effects of environmental concentrations of the antiepileptic drug carbamazepine on biomarkers and cAMP-mediated cell signaling in the mussel *Mytilus galloprovincialis*. Aquat Toxicol 94:177–185

McEneff G, Barron L, Kelleher B, Paull B, Quinn B (2014) A year-long study of the spatial occurrence and relative distribution of pharmaceutical residues in sewage effluent, receiving marine waters and marine bivalves. Sci Total Environ 476:317–326

Monteiro SC, Boxall AB (2010) Occurrence and fate of human pharmaceuticals in the environment. In: Reviews of environmental contamination and toxicology. Springer, Berlin, pp 53–154

Moreno-González R, Rodríguez-Mozaz S, Huerta B, Barceló D, León VM (2016) Do pharmaceuticals bioaccumulate in marine molluscs and fish from a coastal lagoon? Environ Res 146:282–298

Motohashi N, Ikawa K, Kariya T (1989) GABAB receptors are up-regulated by chronic treatment with lithium or carbamazepine. GABA hypothesis of affective disorders? Eur J Pharmacol 166:95–99

Nödler K, Voutsa D, Licha T (2014) Polar organic micropollutants in the coastal environment of different marine systems. Mar Pollut Bull 85:50–59

Oliveira P, Almeida Â, Calisto V, Esteves VI, Schneider RJ, Wrona FJ, Soares AM, Figueira E, Freitas R (2017) Physiological and biochemical alterations induced in the mussel *Mytilus galloprovincialis* after short and long-term exposure to carbamazepine. Water Res 117:102–114

Powell EN, Cummins H (1985) Are molluscan maximum life spans determined by long-term cycles in benthic communities? Oecologia 67:77–182

Prichard E, Granek EF (2016) Effects of pharmaceuticals and personal care products on marine organisms: from single-species studies to an ecosystem-based approach. Environ Sci Pollut Res 23:22365–22384

Regoli F, Giuliani ME (2014) Oxidative pathways of chemical toxicity and oxidative stress biomarkers in marine organisms. Mar Environ Res 93:106–117

Rodrigues J, Albino S, Silva S, Cravo A, Cardoso VV, Benoliel MJ, Almeida CM (2019) Development of a multiresidue method for the determination of 24 pharmaceuticals in clams by QuEChERS and liquid chromatography-triple quadrupole tandem mass spectrometry. Food Anal Methods 12:838–851

Sayre LM, Perry G, Smith MA (2008) Oxidative stress and neurotoxicity. Chem Res Toxicol 21:172–188

Solé M, Shaw JP, Frickers PE, Readman JW, Hutchinson TH (2010) Effects on feeding rate and biomarker responses of marine mussels experimentally exposed to propranolol and acetaminophen. Anal Bioanal Chem 396:649–656

Sousa JCG, Ribeiro AR, Barbosa MO, Pereira MFR, Silva AMT (2018) A review on environmental monitoring of water organic pollutants identified by EU guidelines. J Hazard Mater 344:146–162

Trombini C, Hampel M, Blasco J (2019) Assessing the effect of human pharmaceuticals (carbamazepine, diclofenac and ibuprofen) on the marine clam *Ruditapes philippinarum*: an integrative and multibiomarker approach. Aquat Toxicol 208:146–156

Tsiaka P, Tsarpali V, Ntaikou I, Kostopoulou MN, Lyberatos G, Dailianis S (2013) Carbamazepine-mediated pro-oxidant effects on the unicellular marine algal species *Dunaliella tertiolecta* and the hemocytes of mussel *Mytilus galloprovincialis*. Ecotoxicology 22:1208–1220

Viarengo A, Canesi L (1991) Mussels as biological indicators of pollution. Aquaculture 94:225–243

Voultsiadou E, Koutsoubas D, Achparaki M (2010) Bivalve mollusc exploitation in Mediterranean coastal communities: an historical approach. J Biol Res 13:35

Wille K, Kiebooms JA, Claessens M, Rappé K, Bussche JV, Noppe H, Van Praet N, De Wulf E, Van Caeter P, Janssen CR (2011) Development of analytical strategies using U-HPLC-MS/MS and LC-ToF-MS for the quantification of micropollutants in marine organisms. Anal Bioanal Chem 400:1459–1472

Zhang Y, Geißen S-U, Gal C (2008) Carbamazepine and diclofenac: removal in wastewater treatment plants and occurrence in water bodies. Chemosphere 73:1151–1161

Fine Particulate Matter ($PM_{2.5}$) and Chronic Kidney Disease

Yilin Zhang, Dongwei Liu, and Zhangsuo Liu

Contents

Abstract The impact of ambient particulate matter (PM) on public health has become a great global concern, which is especially prominent in developing countries. For health purposes, PM is typically defined by size, with the smaller particles having more health impacts. Particles with a diameter <2.5 μm are called $PM_{2.5}$. Initial research studies have focused on the impact of $PM_{2.5}$ on respiratory and cardiovascular diseases; nevertheless, an increasing number of data suggested that $PM_{2.5}$ may affect every organ system in the human body, and the kidney is of no exception. The kidney is vulnerable to particulate matter because most environmental toxins are concentrated by the kidney during filtration. According to the high morbidity and mortality related to chronic kidney disease, it is necessary to

Dongwei Liu and Zhangsuo Liu contributed equally with all other contributors.

Y. Zhang · D. Liu (✉) · Z. Liu (✉)
Department of Nephrology, The First Affiliated Hospital of Zhengzhou University, Zhengzhou, P. R. China

Research Institute of Nephrology, Zhengzhou University, Zhengzhou, P. R. China

Research Center for Kidney Disease, Zhengzhou, Henan Province, P. R. China

Key Laboratory of Precision Diagnosis and Treatment for Chronic Kidney Disease in Henan Province, Zhengzhou, P. R. China

Core Unit of National Clinical Medical Research Center of Kidney Disease, Zhengzhou, P. R. China
e-mail: yilinzhang17@gmail.com; liu-dongwei@zzu.edu.cn; zhangsuoliu@zzu.edu.cn

P. de Voogt (ed.), *Reviews of Environmental Contamination and Toxicology Volume 254*,
Reviews of Environmental Contamination and Toxicology 254,
https://doi.org/10.1007/398_2020_62

determine the effect of $PM_{2.5}$ on kidney disease and its mechanism that needs to be identified. To understand the current status of $PM_{2.5}$ in the atmosphere and their potential harmful kidney effects in different regions of the world this review article was prepared based on peer-reviewed scientific papers, scientific reports, and database from government organizations published after the year 1998. In this review, we focus on the worldwide epidemiological evidence linking $PM_{2.5}$ with chronic kidney disease and the effect of $PM_{2.5}$ on the chronic kidney disease (CKD) progression. At the same time, we also discuss the possible mechanisms of $PM_{2.5}$ exposure leading to kidney damage, in order to emphasize the contribution of $PM_{2.5}$ to kidney damage. A global database on $PM_{2.5}$ and kidney disease should be developed to provide new ideas for the prevention and treatment of kidney disease.

Keywords Air pollution · Chronic kidney disease · Environmental health · Fine particulate matter · Kidney injury · Urbanization

Abbreviations

CI	Confidence interval
CKD	Chronic kidney disease
HR	Hazard ratio
OR	Odds ratio
$PM_{2.5}$	Fine particulate matter
US EPA	US Environmental Protection Agency
WHO	World Health Organization

1 Introduction

Air pollution is becoming an ever more serious problem due to rapid urban industrialization and modernization (Chen 2007; Power et al. 2018). The World Health Organization (WHO) reports that 90% of the world's population suffers from the polluted environment, while almost seven million deaths from air pollution every year are caused by exposure to fine particles (an annual average $PM_{2.5}$ concentration above 10 μg/m^3) (World Health Organization 2018a, b, c). Air pollutants are composed of gaseous substances, volatile substances, semi-volatile substances, and particulate matter (PM) mixture (Zanobetti et al. 2009). PMs with diameters that are generally 2.5 μm and smaller (e.g., $PM_{2.5}$) have been found to have a greater impact on human health (Brunekreef and Holgate 2002; Costa et al. 2014). More than half of the global population is exposed to very low quality air ($PM_{2.5}$ concentration > 35 μg/m^3) (World Development Indicators 2017). After enrichment in $PM_{2.5}$, the above substances can be deposited in the alveoli through respiration, and can even enter the blood circulation through the gas–blood barrier, thus reaching other tissues and organs, and causing damage to many systems such as respiration, circulation, and so on (Martinelli et al. 2013; Dominici et al. 2006; Zhu et al. 2015; World Health Organization 2018a, b, c).

Over recent years, the epidemiological studies have shown an obvious upward trend in the incidence of CKD. The global prevalence rate is about 11–13% and the incidence of adult CKD in China is around 10.8% (Zhang et al. 2012; Hill et al. 2016). Thus, the prevention and treatment of CKD are becoming vital public health problems faced by the whole world. The presence of some comorbidities such as cardiovascular disease(CVD), hypertension, and diabetes is a potential predictor of the rapid aggravation from CKD to the end-stage renal disease (ESRD). The increasing number of studies have shown that air pollution may be a new risk factor for CKD (Tonelli et al. 2012; Blum et al. 2020; Lin et al. 2020; Wu et al. 2020).

As an important organ involved in hemofiltration and toxin excretion, kidneys can be easily affected by air pollutants in blood, which is why the effect of $PM_{2.5}$ exposure on kidney disease should not be ignored. In this review, we summarized the epidemiological evidence of kidney disease being associated with $PM_{2.5}$ exposure, as well as the findings about the effects of $PM_{2.5}$ on the progression of kidney disease. At the same time, we also discussed the possible mechanisms of kidney injury caused by $PM_{2.5}$ exposure. Based on these findings, we developed a hypothesis that exposure to $PM_{2.5}$ and CKD is interrelated and there are many biological mechanisms involved in this process.

2 Method

We searched on the World Wide Web for combinations of keywords such as fine particulate matter pollution, $PM_{2.5}$, source apportionment of fine PM, constituents of $PM_{2.5}$, $PM_{2.5}$, and chronic kidney disease (CKD), $PM_{2.5}$ and end-stage renal disease, $PM_{2.5}$ and proteinurian in PubMed, Web of Science and Google Scholar. We selected 350 peer-reviewed articles published from 1998 to 2020 containing information on fine PM and related CKD. Among the articles searched, only articles that met the following criteria were included in the review: (1) Geographical location: Journals from all over the world were considered for the literature review; (2) Sample size: Sample size was not considered in the screening process; (3) Study methodology and statistical analysis: Research methods and the associated statistical analysis were not considered during the screening process; (4) Discussion of health effect: health effects using human or animal subjects. For global and country specific variations in fine PM, databases from government organizations like the National Clean Air Program (NCAP) in India, United States Environmental Protection Agency (US EPA) in the USA, European Environment Agency (EEA) for Europe, and World Health Organization (WHO) for global database were also screened (National Clean Air Programme 2019; European Environment Agency 2019; The U.S. Environmental Protection Agency 2012; World Health Organization 2018a, b, c). This review has selected 133 papers, of which $PM_{2.5}$ and CKD related studies were distributed in the following countries (number of papers): China (34),

Taiwan, China (5), India (3), Japan (2), South Korea (1), East Asia (1), USA (20), Canada (5), South Africa (1), United Arab Emirates (1), New Zealand (2), France (3), the UK (4), The Netherlands (2), Germany (1), Spain (1), Denmark (1), Greece (1), Europe (3), Global (8).

3 The Compositions of $PM_{2.5}$

$PM_{2.5}$ is one of the main components of air pollution and a major risk factor for the global disease burden (Cohen et al. 2017; Monn and Becker 1999). $PM_{2.5}$ is not a simple pollutant, but a mixture of many substances (Harrison et al. 2004). It often derives from different outdoor emission sources and also has different chemical compositions. Chemical components include trace heavy metals (Cd, Cr, Cu, Mn, Ni, Pb, V, and Zn), organic compounds (volatile components and polycyclic aromatic hydrocarbons), carbonaceous aerosols (elemental carbon and organic carbon), inorganic mineral dust, sea salt and water-soluble inorganic ions (NO^{3-},$SO4^{2-}$,Cl^{-}, F^{-},NO^{2-},Br^{-},NH^{4+}, Na^{+}, K^{+}, Ca^{2+}, Mg^{2+}) (Tao et al. 2016; Feng et al. 2006; Morakinyo et al. 2016). To be specific, $PM_{2.5}$ can also originate from indoor sources. Particles of indoor origin include components derived from biological sources. Biological components include pollen, microorganisms (fungi, bacterias, and virus) and organic compounds derived from microorganisms (endotoxin, metabolites, toxins, and other microbial fragments), many of which are known allergens (Douwes et al. 2003). The biological aerosol accounts for 5–34% of the indoor pollutants which may trigger the immune response, inflammation response, infectious disease, cancer, and other toxicity (Pollution Issues 2006; Air Quality Direct 2007; Douwes et al. 2000; Fung and Hughson 2003; Samake et al. 2017; Srikanth et al. 2008).

The wide distribution of heavy metals (metallic elements that have a relatively high density compared to water) and carbonaceous aerosols in $PM_{2.5}$ results in severe biological problems (Zhang et al. 2016; Gadi et al. 2019). The combustion elements such as EC (elemental carbon) and combustion sources such as biomass burning (potassium, as the main trace element) have been shown to be strongly associated with mortality, compared to other components (Achilleos et al. 2017). Typical fuel oils contain Iron (Fe), Nickel (Ni), Vanadium (V), and Zinc (Zn) and they are reflected in the composition of fly ash produced from fossil-fuel combustion (Vouk and Piver 1983). The exposure to zinc may interfere with the vasoconstriction and vasodilation that increase the risk of CVD and upregulate the expression of cytokines and stress proteins (Graff et al. 2004). The exposure to vanadium and chromium potentially has a major role in inducing oxidative DNA damage (Sørensen et al. 2005). Sulfate, ammonium, and nitrate carried in $PM_{2.5}$, usually found in coal combustion and vehicle emissions, are the main contributors to the haze in China, which can rapidly dissolve in the well-buffered lining fluids of the respiratory system (Harrison and Yin 2000). Polycyclic aromatic hydrocarbons (PAHs) in $PM_{2.5}$ are formed by incomplete combustion and pyrolysis of organic substances such as coal, oil, natural gas, and wood, which can lead to carcinogenesis and gene mutation (Bandowe et al. 2014; Lundstedt et al. 2007; Yunker et al. 2002). The US Environmental Protection Agency reports that the acute exposure to PAHs may be harmful to human health (The U.S. Environmental Protection Agency 2012). However, the PAH levels in the atmosphere in China are generally higher than those in foreign countries (Farmer et al. 2003; Lee et al. 2005; Sharma et al. 2007; Wang et al. 2015).

In addition, studies from many institutions in different countries have shown that the concentration, characteristics, and toxicity of $PM_{2.5}$ vary with time, season, location, and climate (Bell et al. 2008; Kim et al. 2019; Lee et al. 2019). Fine PM levels and exceedance of national and international standards were several times higher in Asian countries (Mukherjee and Agrawal 2017). The monitoring data of $PM_{2.5}$ in 45 major cities around the world in 2013 showed that, as a result of the rapid expansion of cities and rapid economic growth, the highly polluted megacities were concentrated in east-central China and the Indo-Gangetic Plain, among which the most polluted areas were Delhi in India, Cairo in Egypt, and Tianjin in China with the highest annual average $PM_{2.5}$ concentration (89 to 143 μg/m^3) (Cheng et al. 2016). The satellite data evaluation model from 2004 to 2014 showed that the estimated value of 10-year average $PM_{2.5}$ in the Beijing-Tianjin metropolitan region (including Beijing, Tianjin, and Hebei) was generally no less than 100 μg/m^3. The 10-year average concentration of $PM_{2.5}$ in Sichuan Basin and Yangtze River Delta is generally no less than 85 μg/m^3, and it is no less than 55 μg/m^3 in Pearl River Delta (Ma et al. 2015). $PM_{2.5}$ levels vary greatly in different regions of the same country. It is worth noting that a recent study found that China's average annual $PM_{2.5}$ concentrations have dropped by 30% and 50% overall (Zhai et al. 2019). PM concentrations collected by the newest air quality monitoring network of the Ministry of Environmental Protection in 190 major cities in China between April 2014 and April 2015 show that the concentration of $PM_{2.5}$ has a significant seasonal variation, which is highest in winter and lowest in summer (Zhang and Cao 2015). Although $PM_{2.5}$ levels in China have gradually decreased in recent years, the incidence of CKD is still very high. The reason for this is probably that $PM_{2.5}$ levels are still higher than the WHO guidelines recommend. In addition, the damage to the kidney caused by long-term exposure to $PM_{2.5}$ is not transient, but usually has a sequelae effect.

Mounting evidence implicates the components of $PM_{2.5}$ can translocate from the lungs into the circulation, are filtered and excreted by the kidneys. Although currently limited, data on the link between air pollution and kidney injury or disease in humans and in animal models are also now beginning to emerge.

4 $PM_{2.5}$ and CKD: Epidemiological Studies

A large cohort study of 100, 629 non-CKD Taiwanese residents aged 20 or over found that long-term exposure to environmental $PM_{2.5}$ was associated with an increased risk of CKD developing. For every 10 μg/m^3 increase in $PM_{2.5}$

concentration, the risk of CKD increased by 6% (HR:1.06, 95% CI:1.02, 1.10) (Chan et al. 2018). Of note, in this study a single measurement of eGFR <60 mL/min/1.73 m^2 was used to define CKD while it might be due to acute kidney disease or other diseases; thus, some participants might have been misclassified as having CKD based on this criterion. Currently, a large number of studies have confirmed the epidemiological evidence of the relationship between renal disease and $PM_{2.5}$ exposure. A large cohort of 2,482,737 veterans in the USA revealed that long-term exposure to $PM_{2.5}$ was associated with an increased risk of CKD during a median follow-up of 8.52 years. For every 10 μg/m^3 increase in $PM_{2.5}$ concentration, the risk of developing CKD increased by 27% (HR, 1.27; 95% CI, 1.17 to 1.38) (Bowe et al. 2017, 2018). However, the study included only US veterans who were mostly older, white men. Therefore, the findings might not be generalizable to other populations. Furthermore, a median exposure to $PM_{2.5}$ of the air quality in the USA is 10–11 μg/m^3, which is better than most countries. A study of 71,151 native biopsies from 938 hospitals in China found that higher $PM_{2.5}$ exposure was associated with the risk of membranous nephropathy. In areas with $PM_{2.5}$ > 70 ug/m^3, each 10 μg/m^3 increase in $PM_{2.5}$ concentration was associated with 14% higher odds for membranous nephropathy (odds ratio, 1.14; 95% CI, 1.10 to 1.18) (Xu et al. 2016). There were some limitations of this study. First, the study included only patients from whom renal biopsy samples were taken and not the general population. Furthermore, information on patient residence were limited to the city level, where levels of $PM_{2.5}$ were generally higher than in other regions and might have led to an underestimation of the effect of $PM_{2.5}$. It is of great clinical significance to studying the dose–response relationship between $PM_{2.5}$ and the development and progression of CKD across a wide range of exposure levels. Bowe and his colleagues recently quantitated the global burden of CKD attributable to ambient air pollution in 2016 and found that more than 30% of disability-adjusted life years of CKD were related to $PM_{2.5}$ exposure globally (Bowe et al. 2019). The study used the Global Burden of Disease study data and provided a quantitative analysis of the global burden of CKD attributable to $PM_{2.5.}$ Although its explanations and conclusions based on a raw analysis, the results provided an intuitive way to translate from the relative risk into attributable burden of CKD due to PM pollutants. Although its explanations and conclusions were based on raw analysis, the results provided an intuitive way to translate from the relative risk into attributable burden of CKD due to PM pollutants (Table 1).

There is a strong reason to believe that $PM_{2.5}$ may be a novel environmental risk factor for CKD. Associations between long-term $PM_{2.5}$ exposure and death were modified by many factors, such as other air pollutants like ozone and mercury (Lelieveld et al. 2019; Stojan et al. 2019), meteorological factors like temperature, resultant wind, relative humidity and barometric pressure, epidemiological and social factors like wealth gap between rich and poor (Kioumourtzoglou et al. 2016). However, in the same environment, some people are more likely than others to develop kidney disease. This may be due to susceptibility of the population.

Previous studies suggested that elderly people and children were especially susceptible to the harmful effects of $PM_{2.5}$ exposure (Mehta et al. 2016; Bowe et al. 2017, 2018). These studies may be influenced by the aging of the population

Table 1 Summary of epidemiological studies on the association between fine particulate matter and chronic kidney disease

Study	Type of study	Study population	Study location	Health outcome
Long-Term Exposure to Air Pollution and Increased Risk of Membranous Nephropathy in China (Xu et al. 2016)	Cohort study	General population 2004–2014	China	Long-term exposure to high-level $PM_{2.5}$ is associated with an increased risk of membranous nephropathy (MN)
Particulate Matter Air Pollution and the Risk of Incident CKD and Progression to ESRD (Bowe et al. 2018)	Cohort study	Veterans	U.S.	There is a significant correlation between the exposure to $PM_{2.5}$ and the risk of CKD, ESRD and the declined eGFR
Long-Term Exposure to Ambient Fine Particulate Matter and Renal Function in Older Men: The Veterans Administration Normative Aging Study (Bowe et al. 2019)	Cross-sectional study	Veterans	U.S.	$PM_{2.5}$ exposure was associated with declined eGFR in the elderly within 1 year
Residential proximity to major roadways and renal function (Lue et al. 2013)	Cohort study	Patients with confirmed acute ischemic stroke 1999–2004	U.S.	Living near main road in patients with acute ischemic stroke is associated with declined eGFR
Associations between Long-Term Particulate Matter Exposure and Adult Renal Function in the Taipei Metropolis (Yang et al. 2017)	Cross-sectional study	General population 2007–2009	Taiwan, China	Exposure to PM10 and PMCoarse, instead of $PM_{2.5}$ or $PM_{2.5}$Absorbance is associated with increased risk of CKD and declined eGFR in adult residents of Taipei
Long-Term Exposure to Ambient Fine Particulate Matter and Chronic Kidney Disease: A Cohort Study (Chan et al. 2018)	Cohort study	General population 2001–2014	Taiwan, China	Long-term exposure to $PM_{2.5}$ is associated with increased risk of CKD. Every 10 $\mu g/m^3$ increase in concentration of $PM_{2.5}$ is associated with 6% increase in the risk of CKD

(continued)

Table 1 (continued)

Study	Type of study	Study population	Study location	Health outcome
County-level air quality and the prevalence of diagnosed chronic kidney disease in the US Medicare population (Bragg-Gresham et al. 2018)	Cross-sectional study	General population	U.S.	There is a positive correlation between concentration of $PM_{2.5}$ and diagnosis of CKD in the county-level cities. When concentration of $PM_{2.5}$ increases by 4 μg/m^3, the adjusted prevalence ratios of diagnosed CKD was 1.03 (95%CI:1.02–1.05; $p < 0.001$)
Effects of long-term exposure to CO and $PM_{2.5}$ on microalbuminuria in type 2 diabetes (Chin et al. 2018)	Cohort study	Patients with type 2 diabetes 2003–2012	Taiwan	Exposure to high-level CO and $PM_{2.5}$ can increase albuminuria in patients with type 2 diabetes mellitus (T2DM)
Associations between Ambient Fine Particulate Levels and Disease Activity in Patients with Systemic Lupus Erythematosus (SLE) (Bernatsky et al. 2010)	Cohort study	Patients (women) with SLE 2000–2007	Canada (Montreal)	Short-term changes in air pollution may affect the disease activity of autoimmune rheumatic diseases. Level of $PM_{2.5}$ is associated with anti-dsDNA and cell tube type
Heat and PAHs Emissions in Indoor Kitchen Air and Its Impact on Kidney Dysfunctions among Kitchen Workers in Lucknow, North India (Singh et al. 2016)	Cross-sectional study	Kitchen workers	North India (Lucknow)	Exposure to $PM_{2.5}$, VOC, and PAH from the indoor air and presence of urinary PAHs metabolites may lead to inflammation. Thus, it may lead to microalbuminuria in kitchen workers
Association of Cigarette Smoking with Albuminuria in the United States: The Third National Health and Nutrition Examination Survey (Hogan et al. 2007)	Cross-sectional study	General population	U.S.	Current smoking and passive smoking are associated with the presence of albuminuria in the general US population with hypertension

(continued)

Table 1 (continued)

Study	Type of study	Study population	Study location	Health outcome
High particulate matter 2.5 levels and ambient temperature are associated with acute lung edema in patients with nondialysis Stage 5 chronic kidney disease (Chiu et al. 2018)	Case-crossover study	Patients with stage 5 nondialysis chronic kidney disease	Taiwan, China	A high $PM_{2.5}$ level was associated with an increased risk of acute lung edema
Associations between long-term ambient $PM_{2.5}$ exposure and prevalence of chronic kidney disease in China: a national cross-sectional study (Huang et al. 2019)	Cross-sectional study	General population 2007–2010	China	There is a significant association between long-term $PM_{2.5}$ exposure with broad concentration ranges and increased prevalence of chronic kidney disease in a representative general Chinese population
Estimates of the 2016 global burden of kidney disease attributable to ambient fine particulate matter air pollution (Bowe et al. 2019)	The global burden of disease (GBD) study	194 countries and territories	Global	The 2016 global burden of incident CKD attributable to $PM_{2.5}$ was 6,950,514 (95% uncertainty interval: 5061533–8,914,745). The global toll of CKD attributable to ambient air pollution is significant and identify several endemic geographies where air pollution may be a significant driver of CKD burden. Air pollution may need to be considered in the discussion of the global epidemiology of CKD
Particulate Matter and Albuminuria, Glomerular Filtration Rate, and Incident CKD (Blum et al. 2020)	Cohort study	Participants from the atherosclerosis risk in communities cohort 1996–1998 through 2016	U.S.	Exposure to higher annual average $PM_{2.5}$ concentrations was associated with a higher level of albuminuria and higher risk for incident CKD in a community-based cohort

(continued)

Table 1 (continued)

Study	Type of study	Study population	Study location	Health outcome
Air pollutants and subsequent risk of chronic kidney disease and end-stage renal disease: A population-based cohort study (Lin et al. 2020)	Cohort study	General population	Taiwan, China	The concentrations of air pollutant were classified into four levels based on quartile. Compared with Q1-level $PM_{2.5}$, exposure to the Q4 level was at a 1.74-fold risk of CKD (95% CI = 1.53–1.98) and 1.69-fold risk of ESRD (95% CI = 1.32–2.16). Exposure to particulate was observed to be associated with an increased risk of CKD and ESRD

MN membranous nephropathy, *CKD* chronic kidney disease, *ESRD* end-stage renal disease, *eGFR* estimated glomerular filtration rate, *T2DM* type 2 diabetes mellitus, *SLE* systemic lupus erythematosus, *VOC* volatile organic compound

in Europe and the USA and the mean age of onset of CKD (Risk factor for CKD: age 60 or older) (Inker et al. 2014). In fact, vascular endothelial dysfunction worsens with age, and the vascular endothelium is more sensitive to pollutants and more vulnerable to injury in older people. Also, children and infants are susceptible to harm from inhaling $PM_{2.5}$ because they inhale more air per pound of body weight than do adults – they breathe faster, spend more time outdoors, and have smaller body sizes. In addition, children's immature immune systems may cause them to be more susceptible to $PM_{2.5}$ than healthy adults. Moreover, the gender differences may exist in the effect of $PM_{2.5}$ exposure on CKD. A study of 47,204 Chinese adults aged 18 years or older found that every 10 μg/m^3 increase in $PM_{2.5}$ was positively associated with the prevalence of chronic kidney disease (odds ratio [OR] 1.33, 95% CI 1.25–1.41, $p < 0.001$). Sex-stratified analyses showed that risk in men (OR 1.42, 95% CI 1.29–1.57) was slightly higher than in women (1.26, 1.17–1.36) (Li et al. 2020). The above-mentioned analyses cannot be generalized. Susceptibility of population is also affected by differences in age and genetic factors in various regions of the world. Therefore, more global longitudinal studies are needed in the future to directly compare the effects of $PM_{2.5}$ on CKD.

5 $PM_{2.5}$ and CKD: Progression and Prevention

In addition to increasing risk of kidney disease, exposure to $PM_{2.5}$ has also been associated with an increased risk of CKD progression. A prospective cohort study of 669 veterans in the Boston area from the Veterans Affairs Normative Aging Study

Institution revealed that every 2.1 μg/m^3 increase in the concentration of $PM_{2.5}$ results in 1.73 m^2 decreases in eGFR (95%CI:2.99 to 0.76) and 0.60 mL/min/1.73m^2 decrease in renal function per year (95%CI:0.79 to 0.40). It should be emphasized that the exposure level of $PM_{2.5}$ in the study population met the environmental air quality requirements of 13.5% in the USA (Standard for 13.0 μg/m^3). Thus, it can be speculated that the higher levels of $PM_{2.5}$ exposure can significantly increase the rate of the renal dysfunction in the exposed patients (Mehta et al. 2016). Other studies have shown that for every 10 mg/m^3 increase of $PM_{2.5}$ concentration, the risk of eGFR decline $\leq$30% increases by 28% (HR, 1.28; 95% CI:1.18 to 1.39), and the risk of ESRD increases by 26% (HR, 1.26;95% CI:1.17 to 1.35) (Bowe et al. 2017, 2018). Road traffic emissions are one of the main sources of $PM_{2.5}$, which has extensive cyto- and genotoxic effects, which in turn can affect the physiology, development, and mortality of a broad subset of the species that encounter roads (Pérez et al. 2010; Leonard and Hochuli 2017). A study of 1,103 patients with acute ischemic stroke in the Boston-area between 1999 and 2004 showed that eGFR levels decreased in patients living closer to major roadway, indicating a decline in renal function. The eGFR levels of patients who live 50 m away from the main road are 3.9 mL/min/1.73m^2 lower than in those who live 1,000 m away (95%CI: 1.0–6.7; $P = 0.007$). The decrease of renal function in patients living 50 m away from the main road is equal to the decrease of natural physiological renal function with the increase of age by 4 years (Lue et al. 2013).

A cross-sectional study of 317 patients with acute pulmonary edema complicated with stage 5 nondialysis chronic nephropathy (CKD 5-ND) in Taiwan found that a high $PM_{2.5}$ level was associated with an increased risk of acute lung edema in patients with CKD 5-ND. High ambient temperature in hot seasons and low ambient temperature in cold seasons were also associated with increased risk (Chiu et al. 2018). There is an approximate linear correlation between the relative changes of albuminuria over a one-year interval and the risk of CKD; and the presence of albuminuria may be a predictor of the CKD aggravation (Sumida et al. 2017). A study of 812 patients with T2DM in Taiwan found that the albumin-to-creatinine ratio (ACR) in patients with exposure to high-level $PM_{2.5}$ increased 3.96 m g/g, while in patients with low-level it increased 3.17 mg/g, indicating that the exposure to higher levels of $PM_{2.5}$ and CO can increase albuminuria in patients with T2DM (Chin et al. 2018). Another clinical cohort study of patients with SLE living on Montreal Island showed that the levels of $PM_{2.5}$ exposure are associated with the anti-dsDNA and the tubular cell type, reflecting severe renal inflammation (Bernatsky et al. 2010). It has been shown that kitchen fumes contain high concentrations of $PM_{2.5}$. By comparing the levels of exposure to $PM_{2.5}$, VOC and PAH from the working environment and burden of these toxic substances on the body in 94 chefs and controls, cross-sectional study revealed that $PM_{2.5}$ may result in the microalbuminuria in kitchen workers (Singh et al. 2016). Smoking is also an important source of indoor $PM_{2.5.}$ A cross-sectional analysis of 15,179 participants in the third National Health and Nutrition Survey in the USA found that in passive smokers, who were grouped by a quartile of serum cotinine levels, the risk of microalbuminuria in highest group was 1.41 times higher than that in the lowest group (Hogan et al. 2007).

Although there is some evidence that CKD progression is associated with $PM_{2.5}$ pollution, it is manifested in an increased risk of progression to ESRD, increase in albuminuria, decrease in eGFR levels, and the emergence of complications such as pulmonary edema. Current data were based on a few cohort studies and were limited to some countries, lacking globally large-scale and long-term cohort studies.

6 $PM_{2.5}$ and CKD: Mechanisms

The toxic mechanisms underlying the relationship between $PM_{2.5}$ and CKD still remain unclear. A large amount of researches from clinical and animal experiments have confirmed the association between exposure to $PM_{2.5}$ and CVD (Combes and Franchineau 2019). Meanwhile, various risk factors of CVD, such as smoking, obesity, hypertension, and diabetes mellitus, are still the risk factors of CKD (Gansevoort et al. 2013; Parikh et al. 2006; Papademetriou et al. 2016). A current study shows that $PM_{2.5}$ can easily enter the circulatory system through alveolar epithelial cells and can eventually damage the cardiovascular system, thereby causing damage to the kidneys (Milojevic et al. 2014). Therefore, we speculate that part of the pathogenic mechanism of the $PM_{2.5}$-related CKD may be similar to the $PM_{2.5}$-related CVD (Fig. 1).

Oxidative Stress Numerous studies have shown that the oxidative stress induced by $PM_{2.5}$ has an important role in the $PM_{2.5}$-mediated toxicity (Li et al. 2008; Weichenthal et al. 2016; Crobeddu et al. 2017). iRHOM2(RHBDF2), a member of the rhomboid family with proteolytic activity, has been explored in the pathogenic mechanism of CKD (Liu et al. 2018; Herrlich and Kefaloyianni 2018). It has been observed that $PM_{2.5}$ promotes the production of ROS and iRhom2 in mice exposed to $PM_{2.5}$, thus resulting in the blocking of Nrf2 signaling pathway, the activation of JNK MAPK signaling pathway and the damage of podocyte, and eventually in the chronic renal injury (Xu et al. 2018; Ge et al. 2018). It has also been observed that the production of ROS induced by $PM_{2.5}$ inhibits the Nrf2 signaling pathway and the activity of downstream genes SOD1, Gclc and GSR in human embryonic stem cells. Meanwhile, N-acetylcysteine (NAC) can restore partial antioxidant stress by scavenging ROS (Jin et al. 2019). NOX4, which is a member of NADPH oxidase family, is considered as the main source of ROS production in the kidneys (Gorin et al. 2005; He et al. 2016). After the intratracheal exposure to the $PM_{2.5}$, the expression of NOX4 and the content of H_2O_2 in the kidneys (the mainly direct metabolite of NOX4) significantly increase while the expression of antioxidant enzyme SOD-1 significantly decreases in the BALB/c mice. Furthermore, the renin–angiotensin system (RAS) in mice can be activated by the $PM_{2.5}$ exposure. It has been reported that Ang-II is one of the most important stimuli to promote the oxidative stress, showing that the renal damage by $PM_{2.5}$ may be induced via activation of RAS, which in turn induces oxidative stress and inflammation (Zhang et al. 2018; Chabrashvili et al. 2003).

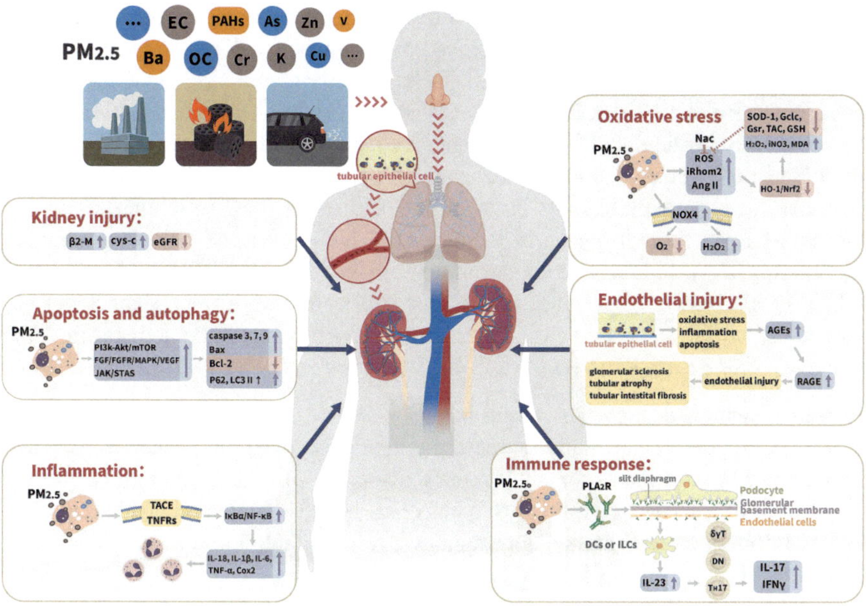

Fig. 1 Plausible biological pathways linking fine particulate matter exposure with chronic kidney disease (CKD). The 6 generalized intermediary pathways and the subsequent specific biological responses that could be capable of instigating CKD events are shown. *eGFR* estimated glomerular filtration rate, *PI3K* phosphatidylinositol-3-kinase, *Akt* protein kinase B, *mTOR* mammalian target of rapamycin, *FGF* fibroblast growth factor, *FGFRs* fibroblast growth factor receptors, *MAPK* mitogen-activated protein kinase, *VEGF* vascular endothelial growth factor, *JAK* Janus kinase, *STAS* spread through air spaces, *Bax* Bcl-2 associated X, *Bcl-2* B cell lymphoma-2, *LC3* light chain 3, *TNFRs* tumor necrosis factor receptors, *TACE* TNF-α converting enzyme, *NF-κB* nuclear factor kappa-B, *IκB* inhibitor of NF-κB, *IL-18* Interleukin-18, *COX2* cyclooxygenase 2, *ROS* reactive oxygen species, *iRhom2* inactive rhomboid protein 2, *Ang II* angiotensin II, *SOD* seminal plasma enzymes, *GCLC* glutamate-cysteine ligase catalytic, *GSR* glutathione reductase, *TAC* total antioxidant capacity, *GSH* glutathione, *MDA* malondialdehyde, *Nrf2* nuclear factor-erythroid 2-related factor 2, *RAGE* advanced glycation end, *IFN* interferon, *PLA2R* phospholipase A2 receptor

Inflammatory Response As mentioned above, $PM_{2.5}$ has an important role in the unhealthy effect induced by the inflammatory response (Bekki et al. 2016; Bo et al. 2016). In the type 1 DM rat model induced by STZ, the levels of the cytokines involved in the inflammatory response including IL-6 and fibrinogen significantly increase after exposure to low concentrations (close to the current air quality standards) and low-enriched $PM_{2.5.}$ Also, the histopathologic studies show that exposure to $PM_{2.5}$ results in glomerulosclerosis and proximal tubular injury (Yan et al. 2014). $PM_{2.5}$ promotes the activation of inflammatory associated TNF-α invertase/TNF-α receptor (TACE/TNFRs) signaling pathway and α/nuclear factor κB (IκBα/NF- κB) signaling pathways in the renal cell lines (including HEK-293, MPC5 and HK-2) from the $PM_{2.5}$ exposed mice (Ge et al. 2018).

Vascular Injury $PM_{2.5}$ may induce endothelial injury through systemic inflammation, oxidative stress, and atherosclerosis, resulting in glomerulosclerosis, tubular atrophy, and tubulointerstitial fibrosis, and eventually chronic kidney diseases (Rui et al. 2016; Bo et al. 2016; Liang et al. 2019). Previous studies have shown that the diesel exhaust particulate (DEP) is the main source of $PM_{2.5}$ and ultrafine particles found in cities (Cho et al. 2004; Pérez et al. 2008). Exposure to DEP aggravates renal vascular injury in the CKD rats induced by adenine, resulting in tubular dilatation, tubular necrosis, tubular type, and decreased renal blood flow (RBF) (Nemmar et al. 2009). AngII can induce endothelial injury by activating NADPH oxidase system (Niazi et al. 2017). After intratracheal instillation of $PM_{2.5}$, increasing levels of circulating AngII are observed in rats, which may be due to the activation of the branches of IRE1 α/XBP1s on the unfolded protein-reactive (UPR) (Xu et al. 2017).

Immune Response Previous studies have shown that environmental pollution may break the immune balance and increase the incidence of chronic diseases (Lee and Lawrence 2018). Some of the autoimmune responses have an important role in the occurrence of a series of nephritis (Clynes et al. 1998; Chang et al. 2011; Lau et al. 2010). Idiopathic membranous nephropathy is also an autoimmune disease characterized by the formation of circulating autoantibodies against secretory phospholipase A2 receptor (PLA2R) (Beck Jr. et al. 2009). Existing studies have shown that exposure to fine particles can promote the formation of autoantibodies and immune complexes (Brown et al. 2003, 2004; Pfau et al. 2004). The deposition of IgG immune complexes has been found in the kidneys of C57Bl/6 mice after exposure to fine particles, while the presence of abnormal glomeruli and tubules suggested the occurrence of glomerulonephritis (Pfau et al. 2008).

Cell Apoptosis and Autophagy Cell apoptosis and autophagy can be induced by $PM_{2.5}$-exposure (Wang et al. 2017a, b; Ding et al. 2017). The activation of PI3K-Akt/mTOR signaling pathway, FGF/FGFR/MAPK/VEGF signaling pathway, and JAK-STAT signaling pathway can be promoted by $PM_{2.5}$, which induces the cell apoptosis and autophagy (Zheng et al. 2017; Wang et al. 2016). A recent study has revealed that the sections of renal tissue from the C57BL/6 mice and the 293 T cells exposed to the Arsenic(As) or SO2 show severe diffuse sclerosing glomerulonephritis. Meanwhile, it has also been found the key members of Caspase family, and the expression of Bax, the indicator of Caspase Pathway, are up-regulated in the 293 T cells, while the expression of Bcl-2 is down-regulated (Ji et al. 2019). It has been reported that a large amount of Cd^{2+} accumulation is found in the kidneys of residents living near industrial areas. Cd^{2+} can significantly reduce the survival rate of renal tubular epithelial cells (HK2) and have a key role in renal injury (Kuang 2018; Thomaidis et al. 2003). The increase of LC3-II and the decrease of p62, which are both markers of cell autophagy, can be observed in the renal tubular ductal epithelial cells (NRK-52E) from the rats exposed to Cadmium for 1 h. This suggests that the short-term exposure to cadmium can rapidly activate cell autophagy and thereby counteract the damage (Lee et al. 2017).

Omics Currently, omics has been widely used in clinical research and it has become a new tool for exploration of the biomarkers in renal disease. After the intratracheal exposure to $PM_{2.5}$, a metabolomics test of the urine shows a significant increase in levels of cresol sulfate and p-cresol glucuronide of the uremic toxins from the exposed rats. Both of these are derivatives of p-cresol in phenylalanine metabolites. Furthermore, the levels of glucuronide acidification of acetaminophen decrease, indicating that $PM_{2.5}$ inhibits the kidney detoxification (Zhang et al. 2017). A global analysis of the circulating miRNA genome from people exposed to traffic-related air pollutants has shown that the expression of miR-148a-3p, which is highly expressed in the kidneys, increases after the exposure to higher levels of pollution. Also, the levels of miR-148b-3p in serum have been described as one of the potential non-invasive biomarkers for the diagnosis of IgA nephropathy (Krauskopf et al. 2018).

Podocyte Injury Recent studies have revealed that exposure to $PM_{2.5}$ can mediate the podocyte injury through a variety of mechanisms, which in turn lead to renal injury (Santana et al. 2016; Yang et al. 2018). As mentioned earlier, the PAHs in $PM_{2.5}$ can lead to carcinogenesis and mutagenicity, and the content of PAHs in the air of China is generally higher compared to foreign countries (Bandowe et al. 2014; Lundstedt et al. 2007; Yunker et al. 2002). The highest content of PAHs of $PM_{2.5}$ was benzo fluoranthene (BbF) observed in the atmosphere in Zhengzhou City (Wang et al. 2017a, b). Based on this, we established a mouse model of co-culture glomerular podocytes and alveolar epithelial cells, and found that the expression of Nephrin protein and Podocin protein decreased and the expression of Desmin protein increased with the increase of the concentration and exposure time to BbF ($P < 0.05$), indicating increasing of podocyte injury. Also, the number of autophagy corpuscles decreased and the expression of Beclin-1 protein and LC3B protein decreased ($P < 0.05$) with the increase of the concentration and time of the exposure, suggesting the decreased levels of autophagy in podocytes. Therefore, we speculate that the inhibition of autophagy pathway may be one of the mechanisms underlying glomerular podocyte injury induced by PAHs of $PM_{2.5}$, which may lead to renal disease. The specific mechanism needs to be further clarified (Zhang et al. 2020).

Toxic exploration of $PM_{2.5}$ is an extremely important issue. Oxidative stress, inflammation, endothelial injury, apoptosis, autophagy, and immune response are the main potential mechanisms in $PM_{2.5}$-induced kidney disease progression. The research findings of in vitro cell and in vivo animal investigations have provided vital insights into the mechanisms of $PM_{2.5}$ exposure in kidney disease progression. Better understandings of the mechanisms associated with $PM_{2.5}$ will allow the development of new strategies to decrease the harmful effects of $PM_{2.5}$ on the pathogenesis of various diseases. Despite all accumulating data regarding pollution and its adverse health effects, there are still gaps that have to be recognized. The pathogenesis is still not fully understood. These studies were almost based on animal models. In the future, we need to characterize which components of $PM_{2.5}$ are primarily responsible for specific disease processes. Moreover, human subject researches need to be more developed (Table 2).

Table 2 Summary of in vivo and in vitro studies on kidney effects of $PM_{2.5}$

Study	Study type	Test subjects/cells	PM2.5 dosage	Mechanism of toxicology
Oxidative stress				
Activated iRhom2 drives prolonged PM2.5 exposure-triggered renal injury in Nrf2-defective mice (Xu et al. 2018)	In vivo and in vitro	C57BL/6 mice and HEK-293 cell and mouse podocyte	115 ± 1.5 lg/m^3 for 6 h/day, 5 times/week (mice) and 0, 50, 100, and 200 lg/mL $PM_{2.5}$ mass for 24 h (cell)	Long-term $PM_{2.5}$ exposure causes chronic renal injury by up-regulation of iRhom2/TACE/TNF-a axis in kidney-resident macrophages
N-acetylcysteine attenuates PM-induced apoptosis by ROS-mediated Nrf2 pathway in human embryonic stem cell (Jin et al. 2019)	In vitro	Human embryonic stem cells (hESCs)	0, 1.5, 6, and 24 μg/mL of $PM_{2.5}$ in corresponding media	$PM_{2.5}$ present strong toxic to hESCs, and induce cell apoptosis by ROS-mediated signaling pathway. Interestingly, NAC can decrease intracellular ROS level and protect hESCs against $PM_{2.5}$-induced cell apoptosis partially through activation of the Nrf2 pathway
The Kidney Injury Induced by Short-Term PM Exposure and the Prophylactic Treatment of Essential Oils in BALB/c Mice (Zhang et al. 2018)	In vivo	BALB/c mice	Intratracheal instillation of 50 μL aqueous $PM_{2.5}$ suspensions (0.5 mg $PM_{2.5}$ in sterile saline) on day 0 and day 2	Short-term exposure to $PM_{2.5}$ induced acute kidney damage associated with oxidative stress and inflammatory response. The activation of RAS might be the pivotal upstream of oxidative stress and inflammatory response induced by $PM_{2.5}$ exposure
Inflammation				
$PM_{2.5}$ collected in China causes inflammatory and oxidative stress responses in macrophages through the multiple pathways (Bekki et al. 2016)	In vitro	Mouse macrophage cell line RAW264.7 and wild-type, (TLR2−/−), (TLR4−/−), (MyD88−/−) mice on a BALB/c	Exposed for 3 h to $PM_{2.5}$ at the dose of 30 g/mL	Future study will be required to investigate the role of macrophages and the contribution of the endotoxin pathway and oxidative stress on the exacerbation of the Th2 immune reaction by exposure to $PM_{2.5}$

(continued)

Table 2 (continued)

Study	Study type	Test subjects/cells	PM2.5 dosage	Mechanism of toxicology
Subchronic effects of inhaled ambient particulate matter on glucose homeostasis and target organ damage in a type 1 diabetic rat model (Yan et al. 2014)	In vivo	Sprague–Dawley (SD) rats (BioLasco Taiwan)	The average concentrations (mean [SD]) of $PM_{2.5}$ were 13.30 [8.65]μ g/m3, for 24 h/day, 7 days/week, for a total of 16 weeks (rats)	PM caused focal myocarditis, aortic medial thickness, advanced glomerulosclerosis, and accentuation of tubular damage of the kidney. PM exposure might induce the macro- and micro-vascular complications in DM through chronic hyperglycemia and systemic inflammation
iRhom2 loss alleviates renal injury in long-term $PM_{2.5}$-exposed mice by suppression of inflammation and oxidative stress (Ge et al. 2018)	In vivo and in vitro	C57BL/6 mice and HEK-293 cell and mouse RAW264.7 macrophages and HK-2 and MPC5 and BMDM cells	150.1 ± 2.5 μg/m3, flow rate of 65 L/min for 6 h/day, 5 times a week (mice)	iRhom2−/− mice exhibited reduced inflammatory response, as evidenced by the reduction of interleukin 1β (IL-1β), IL-6, tumor necrosis factor-α (TNF-α) and IL-18 in kidney samples, which might be, at least partly, through inactivating TNF-α converting enzyme/TNF-α receptors (TACE/TNFRs) and inhibitor of α/nuclear factor κ B (IκBα/NF-κB) signaling pathways
Endothelial injury				
$PM_{2.5}$-induced oxidative stress increases adhesion molecules expression in human endothelial cells through the ERK/AKT/NF-κB dependent pathway (Rui et al. 2016)	In vitro	A human umbilical vein cell line (EA. hy926) cells	25, 50, 100, or 200 μg/mL for 1, 3, 6, 12, or 24 h	$PM_{2.5}$-induced ROS may function as signaling molecules triggering ICAM-1 and VCAM-1 expressions through activating the ERK/AKT/NF-κB-dependent pathway, and further promoting monocyte adhesion to endothelial cells

(continued)

Table 2 (continued)

Study	Study type	Test subjects/cells	PM2.5 dosage	Mechanism of toxicology
Effect of Vitamin E and Omega-3 Fatty Acids on Protecting Ambient $PM_{2.5}$-Induced Inflammatory Response and Oxidative Stress in Vascular Endothelial Cells (Bo et al. 2016)	In vitro	Human umbilical vein endothelial cells (HUVECs)	50 μg/mL	Inflammation and oxidative stress might be parts of the mechanisms linking $PM_{2.5}$ to vascular endothelial injury
Repeat dose exposure of PM triggers the disseminated intravascular coagulation (DIC) in SD rats (Liang et al. 2019).	In vivo	Specific pathogen-free (SPF) male Sprague–Dawley (SD) rats	0, 1.8, 5.4, and 16.2 mg/kg bw every 3 days for 30 days	$PM_{2.5}$ could induce inflammatory response, vascular endothelial injury, and prothrombotic state, eventually resulted in DIC
Diesel exhaust particles in the lung aggravate experimental acute renal failure (Nemmar et al. 2009).	In vivo	Male Wistar rats	0.5 or 1 mg/kg (DEP)	The presence of DEP in the lung aggravated the renal, pulmonary, and systemic effects of CP-induced ARF
IRE1alpha/XBP1s branch of UPR links HIF1alpha activation to mediate ANGII-dependent endothelial dysfunction under particulate matter (PM) 2.5 exposure (Xu et al. 2017).	In vivo and in vitro	Male Sprague–Dawley (SD) rats and human umbilical vein endothelial cells (HUVECs)	1.5 mg/kg weight, every other day for 3 times (rats) and 6.25, 12.5, and 25 μg/mL for 24 h (cell)	$PM_{2.5}$ exposure instigates endoplasmic reticulum instability, leading to the induction of IRE1α/XBP1s branch of UPR and links HIF1α transactivation to mediate ANGII-dependent endothelial dysfunction

(continued)

Table 2 (continued)

Study	Study type	Test subjects/cells	PM2.5 dosage	Mechanism of toxicology
Immune response				
Silica accelerated systemic autoimmune disease in lupus-prone New Zealand mixed mice (Brown et al. 2003).	In vivo	New Zealand mixed (NZM 2410) mice	30 μL saline suspensions of 1 mg crystalline silica, 2 instillations 2 weeks apart	Silica is able to exacerbate the development of autoimmune disease in a genetically susceptible mouse model. The disease acceleration involved increases in proteinuria, autoantibody levels, circulating immune complexes, immune complex deposition, and complement C3 deposition within the kidney
Silica-exposed mice generate autoantibodies to apoptotic cells (Pfau et al. 2004)	In vivo	New Zealand mixed (NZM 2410) mice	30 μL saline suspensions of 1 mg crystalline silica, 2 instillations 2 weeks apart	Silica-induced apoptosis may exacerbate autoimmune responses by exposing antigenic epitopes to the immune system
Immunoglobulin and lymphocyte responses following silica exposure in New Zealand mixed mice (Brown et al. 2004)	In vivo	New Zealand mixed (NZM 2410) mice	30 μL saline suspensions of 1 mg crystalline silica, 2 instillations 2 weeks apart	The silica-induced alterations in immunoglobulin levels, increased TNF-α, increased B1a B cells and CD4+ T cells, with decreased regulatory T cells may provide an environment that allows for increased autoreactivity
Asbestos-induced autoimmunity in C57BL/6 mice (Pfau et al. 2008)	In vivo	C57Bl/6 mice	2 doses × 60 μ g/mouse of amphibole asbestos (tremolite)	Tremolite asbestos led to the production of antinuclear antibodies, immune complex deposition in the kidneys

(continued)

Table 2 (continued)

Study	Study type	Test subjects/cells	PM2.5 dosage	Mechanism of toxicology
Apoptosis and autophagy				
$PM_{2.5}$ induced apoptosis in endothelial cell through the activation of the p53-bax-caspase pathway (Wang et al. 2017b)	In vitro	Human umbilical vein endothelial cells (HUVECs)	0.2, 1, 5, 25 μg/mL for 48 or 72 h	Induction of EC apoptosis is an important mechanism by which ambient $PM_{2.5}$ exposure poses adverse effects on the cardiovascular system
ROS-AKT-mTOR axis mediates autophagy of human umbilical vein endothelial cells induced by cooking oil fumes-derived fine particulate matters in vitro (Ding et al. 2017)	In vitro	Human umbilical vein endothelial cells (HUVECs)	0, 25, 50, 100 μg/mL of COF-derived $PM_{2.5}$, 100 μg/mL COF-derived $PM_{2.5}$ + 10 mmol/L NAC, and 10 mmol/L NAC for 12, 24, and 36 h, respectively	Cooking oil fumes-derived $PM_{2.5}$ (COFs-derived $PM_{2.5}$) exposure can induce oxidative stress and cytotoxic effects. ROS-AKT-mTOR axis plays a critical role in HUVECs autophagy induced by COFs-derived $PM_{2.5}$
Signal Transductions of BEAS-2B Cells in Response to Carcinogenic $PM_{2.5}$ Exposure Based on a Microfluidic System (Zheng et al. 2017)	In vitro	BEAS-2B cells		The ambient fine particles could uptake into the cells by pinocytosis, mainly promoting the PI3K-Akt pathway, FGF/FGFR/MAPK/VEGF signaling, and JAK-STAT pathway
The ER stress regulator Bip mediates cadmium-induced autophagy and neuronal senescence (Wang et al. 2016)	In vitro	The rat pheochromocytoma (PC12) cell line	0, 5, 10, and 20 μM Cd for 24 h	Autophagy regulated by Bip expression after ER stress suppressed Cd-induced neuronal senescence

(continued)

Table 2 (continued)

Study	Study type	Test subjects/cells	PM2.5 dosage	Mechanism of toxicology
Arsenic and sulfur dioxide co-exposure induces renal injury via activation of the NF-kappa-B and caspase signaling pathway (Ji et al. 2019)	In vivo and in vitro	Specific pathogen-free C57BL/6 mice (4 weeks; male) and 293-T cell line (human embryonic kidney cells)	Mice of group As and As+SO2: 5 mg/L As for 60 days. Mice of group SO2 and As +SO2 5 mg/m^3 SO2 for 6 h/day (from days 31 to 60) and the ratio of SO32- and HSO31- in solution is 3:1 (mM: mM). The molecular formula of arsenic is sodium arsenite (cell)	Co-exposure caused more severe diffuse sclerosing glomerulonephritis than As and SO2 exposure alone. Apoptosis was aggravated by co-exposure of As and SO2 in 293T cells. As and SO2 cause cell toxicity through increasing oxidative stress, which was increased by co-exposure
Characterization of lead, cadmium, arsenic, and nickel in $PM_{2.5}$ particles in the Athens atmosphere, Greece (Thomaidis et al. 2003)	In vitro	Pb, Cd, As, and Ni in $PM_{2.5}$ particles		It was found that Pb, As, and Ni have common sources, which could be vehicles emissions/ oil combustion and resuspended road dust. Cd and a portion of As originate from industrial activities. Pb exhibited higher values during the winter period
Initial autophagic protection switches to disruption of autophagic flux by lysosomal instability during cadmium stress accrual in renal NRK-52E cells (Lee et al. 2017)	In vitro	NRK-52E cell	5, 10, 25, and 50 μM Cd2+ for a period of 0.5 h to maximally 24 h	The renal proximal tubule (PT) is the major target of cadmium (Cd2+) toxicity where Cd2+ causes stress and apoptosis
Histology technology				
Effects of the ambient fine particulate matter ($PM_{2.5}$) exposure on urinary metabolic profiles in rats using UPLC-Q-TOF-MS (Zhang et al. 2017)	In vivo	Male SD rats	$PM_{2.5}$ suspension was injected into the trachea of rats, 2 times/week for 4 weeks	17 potential endogenous metabolites were identified from urinary samples in rats by UPLC-Q-TOF-MS

(continued)

Table 2 (continued)

Study	Study type	Test subjects/cells	PM2.5 dosage	Mechanism of toxicology
The human circulating miRNome reflects multiple organ disease risks in association with short-term exposure to traffic-related air pollution (Krauskopf et al. 2018)	In vivo	Non-smoking participants, either healthy or suffering from ischemic heart disease (IHD) or chronic obstructive pulmonary disease (COPD)	Participants walked about 6 km/2 h (an exposure period), personal ambient air pollution levels of PM were assessed using a real-time condensation particle counter	54 circulating miRNAs to be dose- and pollutant species-dependently associated with $PM_{2.5}$ already after 2 h of exposure. These circulating miRNAs actually reflect the adverse consequences of traffic pollution-induced toxicity in target tissues including the lung, heart, kidney, and brain

iRhom2 inactive rhomboid protein 2, *Nrf2* nuclear factor erythroid-2 related factor 2, *C57BL/6 mice* C57 black 6 mice, *HEK-293 cell* human embryonic kidney 293 cell, *TACE* tumor necrosis factor-α converting enzyme, *TNF-a* tumor necrosis factor-α, *ROS* reactive oxygen species, *hESCs* human embryonic stem cells, *NAC* N-Acetyl-L-cysteine, *BALB/c Mice* BALB/cByJ mice, *RAS* renin–angiotensin system, *RAW264.7* leukemia cells in mouse macrophage, *TLR* toll-like receptors, *MyD88* myeloid differentiation factor 88, *Th2* T helper 2 cell, *SD rats* Sprague–Dawley rats, *HK-2* human renal tubular epithelial cells, *MPC5* mouse podocyte clone 5, *BMDM* bone marrow derived macrophage, *IL* interleukin, *TNFRs* tumor necrosis factor receptors, *IκBα* inhibitor of NF-κB α, *NF-κB* nuclear factor kappa-B, *ERK* extracellular regulated protein kinases, *AKT* protein kinase B, *ICAM-1* intercellular cell adhesion molecule-1, *VCAM-1* vascular cell adhesion molecule-1, *HUVECs* human umbilical vein endothelial cells, *DIC* disseminated intravascular coagulation, *SPF* specific pathogen free, *DEP* diesel exhaust particles, *ARF* acute renal failure, *IRE1* inositol-requiring enzyme-1, *XBP1s* spliced X-box binding protein 1, *UPR* unfolded protein response, *HIF1* hypoxic inducible factor 1, *ANGII* angiotensin II, *NZM mice* New Zealand mixed mice, *EC* endothelial cell, *mTOR* mammalian target of rapamycin, *COF* cooking oil fumes, *FGF* fibroblast growth factor, *FGFR* fibroblast growth factor receptor, *MAPK* mitogen-activated protein kinase, *VEGF* vascular endothelial growth factor, *JAK-STAT* JAK: Janus Kinase, *STAT* signal transducer and activator of transcription, *ER* endoplasmic reticulum, *PC* pheochromocytoma, *NRK-52E cells* rat renal tubular epithelial cells, *PT* proximal tubule, *IHD* ischemic heart disease, *COPD* chronic obstructive pulmonary disease, *miRNAs* microRNAs

7 Global Implication of Polices and Measurements

The reduction of $PM_{2.5}$ in general requires concerted action by public authorities, industry, and individuals at national, regional, and even international levels. Since the harmful effect of $PM_{2.5}$ on human health, and especially the kidneys, has been confirmed by many studies, the effective management of $PM_{2.5}$ is necessary to reduce health risks to a minimum. In 2005, WHO announced that the recommended concentration of $PM_{2.5}$ was less than 10 μg/m^3 for an annual average and less than 24 μg/m^3 for a 24-h average. However, these are hard to achieve. Thus, WHO gave a three-tier interim goal (World Health Organization 2006). The Non-Communicable Diseases (NCDs) represent the largest and fastest growing threat to human health (World Health Organization 2014). Traditionally, WHO introduced the association as 4 × 4 model: four main NCDs included cardiovascular diseases, cancer, chronic respiratory diseases, and four main modifiable risk factors included tobacco use, unhealthy diet, physical inactivity, and harmful use of alcohol. The UN Political Declaration of 2018 on NCDs added air pollution as the fifth risk factor, rebranding a more comprehensive "5 × 5" model for NCDs management and control (Renshaw et al. 2019).

Many countries around the world have taken various measures to reduce $PM_{2.5}$. In some relatively clean countries, such as the USA, fine particulate pollution decreased 25% on average across the country between 2010 and 2016 (Clay and Muller 2019). The US administration has taken steps to weaken air quality and climate regulations for this. The US EPA strengthened the nation's air quality standards for fine particle pollution to improve public health protection by revising the primary annual $PM_{2.5}$ standard to 12 μg/m^3 (The U.S. Environmental Protection Agency 2012). Most member states in Europe would reach $PM_{2.5}$ levels close to or even below the WHO guideline value (of 10 μg/m^3), and well below the current EU target value (which will be transformed into a limit value in 2015) of annual 25 μg/m^3. In 2017, Estonia, Finland, and Norway did not report any concentrations above the WHO air quality guidelines for $PM_{2.5}$ by EEA Report (European Environment Agency 2019). In addition to using clean energy, green technology, reduction in emissions, and lower population density, the European Environment Agency's (EEA) European Air Quality Index provides citizens with a tool for checking the air quality in their city and supports public engagement in efforts to reduce air pollution. At the same time, India's air quality does still not show an optimistic picture as one of the most polluted countries in the world. 11 of the 12 cities with the highest levels are located in India when looking at the database's ranking of particulate pollution in cities from WHO (World Health Organization 2018a, b, c). The Indian government launched the National Clean Air Program (NCAP), a five-year action plan to curb air pollution, build a pan-India air quality monitoring network, and improve citizen awareness (National Clean Air Programme 2019). The programme focuses on 102 polluted Indian cities and aims to reduce $PM_{2.5}$ levels by 20–30% over the next 5 years. By 2017, Beijing's $PM_{2.5}$ pollution ranking in the world's cities dropped from the previous 40th to 187th. $PM_{2.5}$ pollution in 62 other cities in China also fell by 30% between 2013 and 2017(Greenstone 2018). Moreover, targeted measures should be taken to prevent the $PM_{2.5}$ pollution, such as using air filtration devices, air anion nebulizers and wearing professional dust masks. Meanwhile, kitchen fumes should be controlled and the indoor smoking should be eliminated. Recently, some scholars have suggested taking Vitamin B as an effective way to resist the $PM_{2.5}$ pollution. They found that Vitamin B can reduce the DNA methylation changes induced by $PM_{2.5}$, and that it has an important role in the inflammation and the oxidative stress (Zhong et al. 2017).

8 Conclusions

Fine particulates are important risk factors for kidney disease, especially in developing countries in which environmental pollution is prevalent. The causes of CKD and the toxic mechanisms of most environmental nephrotoxins remain to be elucidated. This paper discussed a number of large cohort studies and cross-sectional studies that associate $PM_{2.5}$ and CKD, as well as the potential mechanisms of $PM_{2.5}$ exposure in kidney disease progression. As mentioned earlier, CKD cases from countries such as America and Asia are estimated to be attributable to long-term exposure to $PM_{2.5}$, and $PM_{2.5}$ may accelerate the progression of CKD to ESRD. In addition, studies on the pathogenic mechanisms of $PM_{2.5}$ involve some molecular biological networks and cellular signaling pathways. In order to understand the causal relationship between $PM_{2.5}$ exposure and CKD, it is important to study the toxic effects of various components in $PM_{2.5,}$ as well as the effects of temperature, season and regional distribution on them. In terms of mechanisms, most studies have focused on the effects of certain specific components in $PM_{2.5}$ in relation to oxidative stress, inflammatory response, immune response, and other general pathways. Lack of specific mechanisms of action and the pathogenesis still need to be fully elucidated. More detailed longitudinal studies and experimental designs are needed to prove which components of $PM_{2.5}$ are responsible for CKD processes and the dose-response relationship between $PM_{2.5}$ and the development and progression of CKD.

In short, $PM_{2.5}$ is an important risk factor for kidney disease. A database of $PM_{2.5}$ and chronic kidney disease should be globally developed. The data from different regions should be released every year or quarter to provide a new perspective at $PM_{2.5}$. This new angle might help to promote the prevention and control of chronic kidney diseases in various countries around the world. In the exploration of relevant mechanisms, from the epigenetics, genomics, proteomics, metabolomics and other levels exploring the impact of $PM_{2.5}$ on CKD, impact mechanisms might provide experimental basis for the prevention and treatment of CKD.

We believe that via joint efforts of various countries around the world, the environmental pollution might not only be improved, but the prevention and control of CKD could also be raised to a new level.

Acknowledgments This study is supported by grants from National Natural Science Foundation of China-Henan Joint Fund, China (No.U1904146), Comprehensive and digital demonstration platform for clinical evaluation technology of new drugs for major diseases, China (No. 2020ZX09201-009), the Innovation Scientists and Technicians Troop Construction Projects of Henan Province, China (No.182101510002). National Science and Technology Major Project of China, China (No. 2020ZX09201-009).

Author Disclosures No competing interests are reported.

References

Achilleos S, Kioumourtzoglou M-A, Wu C-D, Schwartz JD, Koutrakis P, Papatheodorou SI (2017) Acute effects of fine particulate matter constituents on mortality: a systematic review and meta-regression analysis. Environ Int 109:89–100. https://doi.org/10.1016/j.envint.2017.09.010

Air Quality Direct (2007) Bio-aerosols. http://www.airqualitydirect.com/bio-aerosols.htm

Bandowe BAM, Meusel H, R-j H, Ho K, Cao J, Hoffmann T, Wilcke W (2014) PM2. 5-bound oxygenated PAHs, nitro-PAHs and parent-PAHs from the atmosphere of a Chinese megacity: seasonal variation, sources and cancer risk assessment. Sci Total Environ 473:77–87. https://doi.org/10.1016/j.scitotenv.2013.11.108

Beck LH Jr, Bonegio RG, Lambeau G, Beck DM, Powell DW, Cummins TD, Klein JB, Salant DJ (2009) M-type phospholipase A2 receptor as target antigen in idiopathic membranous nephropathy. N Engl J Med 361:11–21. https://doi.org/10.1056/NEJMoa0810457

Bekki K, Ito T, Yoshida Y, He C, Arashidani K, He M, Sun G, Zeng Y, Sone H, Kunugita N (2016) PM2. 5 collected in China causes inflammatory and oxidative stress responses in macrophages through the multiple pathways. Environ Toxicol Pharmacol 45:362–369. https://doi.org/10.1016/j.etap.2016.06.022

Bell ML, Ebisu K, Peng RD, Walker J, Samet JM, Zeger SL, Dominici F (2008) Seasonal and regional short-term effects of fine particles on hospital admissions in 202 US counties, 1999–2005. Am J Epidemiol 168:1,301–1,310. https://doi.org/10.1093/aje/kwn252

Bernatsky S, Fournier M, Pineau CA, Clarke AE, Vinet E, Smargiassi A (2010) Associations between ambient fine particulate levels and disease activity in patients with systemic lupus erythematosus (SLE). Environ Health Perspect 119:45–49. https://doi.org/10.1289/ehp.1002123

Blum MF, Surapaneni A, Stewart JD, Liao D, Yanosky JD, Whitsel EA, Grams ME (2020) Particulate matter and albuminuria, glomerular filtration rate, and incident CKD. Clin J Am Soc Nephrol 15(3):311–319. https://doi.org/10.2215/CJN.08350719

Bo L, Jiang S, Xie Y, Kan H, Song W, Zhao J (2016) Effect of vitamin E and omega-3 fatty acids on protecting ambient PM2. 5-induced inflammatory response and oxidative stress in vascular endothelial cells. PLoS One 11:e0152216. https://doi.org/10.1371/journal.pone.0152216

Bowe B, Xie Y, Li T, Yan Y, Xian H, Al-Aly Z (2017) Associations of ambient coarse particulate matter, nitrogen dioxide, and carbon monoxide with the risk of kidney disease: a cohort study. Lancet Planet Health 1:e267–e276. https://doi.org/10.1016/S2542-5196(17)30117-1

Bowe B, Xie Y, Li T, Yan Y, Xian H, Al-Aly Z (2018) Particulate matter air pollution and the risk of incident CKD and progression to ESRD. J Am Soc Nephrol 29:218–230. https://doi.org/10.1681/ASN.2017030253

Bowe B, Xie Y, Li T, Yan Y, Xian H, Al-Aly Z (2019) Estimates of the 2016 global burden of kidney disease attributable to ambient fine particulate matter air pollution. BMJ Open 9(5): e022450. https://doi.org/10.1136/bmjopen-2018-022450

Bragg-Gresham J, Morgenstern H, McClellan W, Saydah S, Pavkov M, Williams D et al (2018) County-level air quality and the prevalence of diagnosed chronic kidney disease in the US Medicare population. PLoS One 13(7):e0200612. https://doi.org/10.1371/journal.pone.0200612

Brown J, Archer A, Pfau J, Holian A (2003) Silica accelerated systemic autoimmune disease in lupus-prone New Zealand mixed mice. Clin Exp Immunol 131:415–421. https://doi.org/10.1046/j.1365-2249.2003.02094.x

Brown JM, Pfau JC, Holian A (2004) Immunoglobulin and lymphocyte responses following silica exposure in New Zealand mixed mice. Inhal Toxicol 16:133–139. https://doi.org/10.1080/08958370490270936

Brunekreef B, Holgate ST (2002) Air pollution and health. Lancet 360:1233–1242. https://doi.org/10.1016/S0140-6736(02)11274-8

Chabrashvili T, Kitiyakara C, Blau J, Karber A, Aslam S, Welch WJ, Wilcox CS (2003) Effects of ANG II type 1 and 2 receptors on oxidative stress, renal NADPH oxidase, and SOD expression.

Am J Physiol Regul Integr Compar Physiol 285:R117–R124. https://doi.org/10.1152/ajpregu.00476.2002
Chan T-C, Zhang Z, Lin B-C, Lin C, Deng H-B, Chuang YC, Chan JW, Jiang WK, Tam T, Chang L-y (2018) Long-term exposure to ambient fine particulate matter and chronic kidney disease: a cohort study. Environ Health Perspect 126:107002. https://doi.org/10.1289/EHP3304
Chang A, Henderson SG, Brandt D, Liu N, Guttikonda R, Hsieh C, Kaverina N, TO Utset, Meehan SM, Quigg RJ (2011) In situ B cell-mediated immune responses and tubulointerstitial inflammation in human lupus nephritis. J Immunol 186:1849–1860. https://doi.org/10.4049/jimmunol.1001983
Chen J (2007) Rapid urbanization in China: a real challenge to soil protection and food security. Catena 69:1–15. https://doi.org/10.1016/j.catena.2006.04.019
Cheng Z, Luo L, Wang S, Wang Y, Sharma S, Shimadera H, Wang X, Bressi M, de Miranda RM, Jiang J (2016) Status and characteristics of ambient PM2. 5 pollution in global megacities. Environ Int 89:212–221. https://doi.org/10.1016/j.envint.2016.02.003
Chin W-S, Chang Y-K, Huang L-F, Tsui H-C, Hsu C-C, Guo Y-LL (2018) Effects of long-term exposure to CO and PM2. 5 on microalbuminuria in type 2 diabetes. Int J Hyg Environ Health 221:602–608. https://doi.org/10.1016/j.ijheh.2018.04.009
Chiu P-F, Chang C-H, Wu C-L, Chang T-H, Tsai C-C, Kor C-T, Li J-R, Kuo C-L, Huang C-S, Chu CC (2018) High particulate matter 2.5 levels and ambient temperature are associated with acute lung edema in patients with nondialysis Stage 5 chronic kidney disease. Nephrol Dial Transplant. https://doi.org/10.1093/ndt/gfy144
Cho AK, Di Stefano E, You Y, Rodriguez CE, Schmitz DA, Kumagai Y, Miguel AH, Eiguren-Fernandez A, Kobayashi T, Avol E (2004) Determination of four quinones in diesel exhaust particles, SRM 1649a, and atmospheric PM2. 5 special issue of aerosol science and technology on findings from the fine particulate matter supersites program. Aerosol Sci Tech 38:68–81. https://doi.org/10.1080/02786820390229471
Clay K, Muller NZ (2019) Recent increases in air pollution: evidence and implications for mortality. National Bureau of Economic Research. https://doi.org/10.3386/w26381
Clynes R, Dumitru C, Ravetch JV (1998) Uncoupling of immune complex formation and kidney damage in autoimmune glomerulonephritis. Science 279:1052–1054. https://doi.org/10.1126/science.279.5353.1052
Cohen AJ, Brauer M, Burnett R, Anderson HR, Frostad J, Estep K, Balakrishnan K, Brunekreef B, Dandona L, Dandona R (2017) Estimates and 25-year trends of the global burden of disease attributable to ambient air pollution: an analysis of data from the Global Burden of Diseases Study 2015. Lancet 389:1907–1918. https://doi.org/10.1016/S0140-6736(17)30505-6
Combes A, Franchineau G (2019) Fine particle environmental pollution and cardiovascular diseases. Metabolism 100:153944. https://doi.org/10.1016/j.metabol.2019.07.008
Costa S, Ferreira J, Silveira C, Costa C, Lopes D, Relvas H, Borrego C, Roebeling P, Miranda AI, Paulo Teixeira J (2014) Integrating health on air quality assessment – review report on health risks of two major European outdoor air pollutants: PM and NO2. J Toxicol Environ Health Part B 17:307–340. https://doi.org/10.1080/10937404.2014.946164
Crobeddu B, Aragao-Santiago L, Bui L-C, Boland S, Squiban AB (2017) Oxidative potential of particulate matter 2.5 as predictive indicator of cellular stress. Environ Pollut 230:125–133. https://doi.org/10.1016/j.envpol.2017.06.051
Ding R, Zhang C, Zhu X, Cheng H, Zhu F, Xu Y, Liu Y, Wen L, Cao J (2017) ROS-AKT-mTOR axis mediates autophagy of human umbilical vein endothelial cells induced by cooking oil fumes-derived fine particulate matters in vitro. Free Radic Biol Med 113:452–460. https://doi.org/10.1016/j.freeradbiomed.2017.10.386
Dominici F, Peng RD, Bell ML, Pham L, McDermott A, Zeger SL, Samet JM (2006) Fine particulate air pollution and hospital admission for cardiovascular and respiratory diseases. JAMA 295:1127–1134. https://doi.org/10.1001/jama.295.10.1127
Douwes J, Wouters I, Dubbeld H, van Zwieten L, Steerenberg P, Doekes G, Heederik D (2000) Upper airway inflammation assessed by nasal lavage in compost workers: a relation with

bio-aerosol exposure. Am J Ind Med 37:459–468. https://doi.org/10.1002/(SICI)1097-0274(200005)37:5<459::AID-AJIM2>3.0.CO;2-J

Douwes J, Thorne P, Pearce N, Heederik D (2003) Bioaerosol health effects and exposure assessment: progress and prospects. Ann Occup Hyg 47:187–200. https://doi.org/10.1093/annhyg/meg032

European Environment Agency (2019) Air quality in Europe – 2019 report. https://www.eea.europa.eu/publications/air-quality-in-europe-2019

Farmer PB, Singh R, Kaur B, Sram RJ, Binkova B, Kalina I, Popov TA, Garte S, Taioli E, Gabelova A (2003) Molecular epidemiology studies of carcinogenic environmental pollutants: effects of polycyclic aromatic hydrocarbons (PAHs) in environmental pollution on exogenous and oxidative DNA damage. Mut Res Rev Mut Res 544:397–402. https://doi.org/10.1016/j.mrrev.2003.09.002

Feng J, Chan CK, Fang M, Hu M, He L, Tang X (2006) Characteristics of organic matter in PM2. 5 in Shanghai. Chemosphere 64:1393–1400. https://doi.org/10.1016/j.chemosphere.2005.12.026

Fung F, Hughson WG (2003) Health effects of indoor fungal bioaerosol exposure. Appl Occup Environ Hyg 18:535–544. https://doi.org/10.1080/10473220301451

Gadi R, Sharma SK, Mandal TK (2019) Seasonal variation, source apportionment and source attributed health risk of fine carbonaceous aerosols over National Capital Region, India. Chemosphere 237:124500. https://doi.org/10.1016/j.chemosphere.2019.124500

Gansevoort RT, Correa-Rotter R, Hemmelgarn BR, Jafar TH, Heerspink HJL, Mann JF, Matsushita K, Wen CP (2013) Chronic kidney disease and cardiovascular risk: epidemiology, mechanisms, and prevention. Lancet 382:339–352. https://doi.org/10.1016/S0140-6736(13)60595-4

Ge CX, Xu MX, Qin YT, Gu TT, Lv JX, Wang MX,Wang SJ, Ma YJ, Lou DS, Li Q, Hu LF, Tan J (2018) iRhom2 loss alleviates renal injury in long-term PM2. 5-exposed mice by suppression of inflammation and oxidative stress. Redox Biol 19:147–157. https://doi.org/10.1016/j.redox.2018.08.009

Gorin Y, Block K, Hernandez J, Bhandari B, Wagner B, Barnes JL, Abboud HE (2005) Nox4 NAD (P) H oxidase mediates hypertrophy and fibronectin expression in the diabetic kidney. J Biol Chem 280:39616–39626. https://doi.org/10.1074/jbc.M502412200

Graff DW, Cascio WE, Brackhan JA, Devlin RB (2004) Metal particulate matter components affect gene expression and beat frequency of neonatal rat ventricular myocytes. Environ Health Perspect 112:792–798. https://doi.org/10.1289/ehp.112-1241994

Greenstone M (2018) Four years after declaring war on pollution, China is winning New York Times, A8.

Harrison RM, Yin J (2000) Particulate matter in the atmosphere: which particle properties are important for its effects on health? Sci Total Environ 249:85–101. https://doi.org/10.1016/S0048-9697(99)00513-6

Harrison RM, Jones AM, Lawrence RG (2004) Major component composition of PM10 and PM2. 5 from roadside and urban background sites. Atmos Environ 38:4531–4538. https://doi.org/10.2147/IJNRD.S39739

He T, Xiong J, Nie L, Yu Y, Guan X, Xu X, Xiao T, Yang K, Liu L, Zhang D (2016) Resveratrol inhibits renal interstitial fibrosis in diabetic nephropathy by regulating AMPK/NOX4/ROS pathway. J Mol Med 94:1359–1371. https://doi.org/10.1007/s00109-016-1451-y

Herrlich A, Kefaloyianni E (2018) iRhoms: a potential path to more specific therapeutic targeting of lupus nephritis. https://doi.org/10.1053/j.ajkd.2018.04.010

Hill NR, Fatoba ST, Oke JL, Hirst JA, O'Callaghan CA, Lasserson DS, Hobbs FR (2016) Global prevalence of chronic kidney disease–a systematic review and meta-analysis. PLoS One 11:e0158765. https://doi.org/10.1371/journal.pone.0158765

Hogan SL, Hogan SL, Vupputuri S, Guo X, Cai J, Colindres RE, Heiss G, Coresh J (2007) Association of cigarette smoking with albuminuria in the United States: the third National Health and Nutrition Examination Survey. Ren Fail 29:133–142. https://doi.org/10.1080/08860220601098888

Huang J, Li G, Wang J, Wu S, Guo X, Zhang L (2019) Associations between long-term ambient PM2· 5 exposure and prevalence of chronic kidney disease in China: a national cross-sectional study. Lancet 394:S93. https://doi.org/10.1016/S0140-6736(19)32429-8

Inker LA, Astor BC, Fox CH, Isakova T, Lash JP, Peralta CA, Tamura MK, Feldman HI (2014) KDOQI US commentary on the 2012 KDIGO clinical practice guideline for the evaluation and management of CKD. Am J Kidney Dis 63:713–735. https://doi.org/10.1053/j.ajkd.2014.01.416

Ji P-y, Z-y L, Wang H, Dong J-t, Li X-j, Yi H-l (2019) Arsenic and sulfur dioxide co-exposure induce renal injury via activation of the NF-κB and caspase signaling pathway. Chemosphere 224:280–288. https://doi.org/10.1016/j.chemosphere.2019.02.111

Jin L, Ni J, Tao Y, Weng X, Zhu Y, Yan J, Hu B (2019) N-acetylcysteine attenuates PM2. 5-induced apoptosis by ROS-mediated Nrf2 pathway in human embryonic stem cells. Sci Total Environ 666:713–720. https://doi.org/10.1016/j.scitotenv.2019.02.307

Kim H, Kim H, Lee J-T (2019) Spatial variation in lag structure in the short-term effects of air pollution on mortality in seven major South Korean cities, 2006–2013. Environ Int 125:595–605. https://doi.org/10.1016/j.envint.2018.09.004

Kioumourtzoglou MA, Schwartz J, James P, Dominici F, Zanobetti A (2016) PM2. 5 and mortality in 207 US cities: modification by temperature and city characteristics. Epidemiology (Cambridge, Mass) 27(2):221. https://doi.org/10.1097/EDE.0000000000000422

Krauskopf J, Caiment F, van Veldhoven K, Chadeau-Hyam M, Sinharay R, Chung KF, Cullinan P, Collins P, Barratt B, Kelly FJ (2018) The human circulating miRNome reflects multiple organ disease risks in association with short-term exposure to traffic-related air pollution. Environ Int 113:26–34. https://doi.org/10.1016/j.envint.2018.01.014

Kuang X (2018) A study on the key role of cadmium accumulation induced by PM2. 5 Exposure in renal injury. In: ISEE conference abstracts

Lau KK, Suzuki H, Novak J, Wyatt RJ (2010) Pathogenesis of Henoch-Schönlein purpura nephritis. Pediatr Nephrol 25:19. https://doi.org/10.1007/s00467-009-1230-x

Lee F, Lawrence DA (2018) From infections to anthropogenic inflicted pathologies: Involvement of immune balance. J Toxicol Environ Health Part B 21:24–46. https://doi.org/10.1080/10937404.2017.1412212

Lee RG, Coleman P, Jones JL, Jones KC, Lohmann R (2005) Emission factors and importance of PCDD/Fs, PCBs, PCNs, PAHs and PM10 from the domestic burning of coal and wood in the UK. Environ Sci Technol 39:1436–1447. https://doi.org/10.1021/es048745i

Lee W-K, Probst S, Santoyo-Sánchez M, Al-Hamdani W, Diebels I, von Sivers J-K, Kerek E, Prenner E, Thévenod F (2017) Initial autophagic protection switches to disruption of autophagic flux by lysosomal instability during cadmium stress accrual in renal NRK-52E cells. Arch Toxicol 91:3225–3245. https://doi.org/10.1007/s00204-017-1942-9

Lee CS, Chou C-K, Cheung H, Tsai C-Y, Huang W-R, Huang S-H, Chen M-J, Liao H-T, Wu C-F, Tsao T-M (2019) Seasonal variation of chemical characteristics of fine particulate matter at a high-elevation subtropical forest in East Asia. Environ Pollut 246:668–677. https://doi.org/10.1016/j.envpol.2018.11.033

Lelieveld J, Klingmüller K, Pozzer A, Burnett R, Haines A, Ramanathan V (2019) Effects of fossil fuel and total anthropogenic emission removal on public health and climate. Proc Natl Acad Sci 116:7192–7197. https://doi.org/10.1073/pnas.1819989116

Leonard RJ, Hochuli DF (2017) Exhausting all avenues: why impacts of air pollution should be part of road ecology. Front Ecol Environ 15:443–449. https://doi.org/10.1002/fee.1521

Li N, Xia T, Nel AE (2008) The role of oxidative stress in ambient particulate matter-induced lung diseases and its implications in the toxicity of engineered nanoparticles. Free Radic Biol Med 44:1689–1699. https://doi.org/10.1016/j.freeradbiomed.2008.01.028

Li GX, Huang J, Wang JW, Zhao MH, Liu Y, Guo XB, Wu SW, Zhang LX (2020) Long-Term Exposure to Ambient PM2. 5 and Increased Risk of CKD Prevalence in China. Journal of the American Society of Nephrology. ASN.2020040517. https://doi.org/10.1681/ASN.2020040517

Liang S, Zhao T, Hu H, Shi Y, Xu Q, Miller MR, Duan J, Sun Z (2019) Repeat dose exposure of PM2. 5 triggers the disseminated intravascular coagulation (DIC) in SD rats. Sci Total Environ 663:245–253. https://doi.org/10.1016/j.scitotenv.2019.01.346

Lin SY, Ju SW, Lin CL, Hsu WH, Lin CC, Ting IW, Kao CH (2020) Air pollutants and subsequent risk of chronic kidney disease and end-stage renal disease: a population-based cohort study. Environ Pollut 261:114154. https://doi.org/10.1016/j.envpol.2020.114154

Liu S, Ye L, Tao J, Ge C, Huang L, Yu J (2018) Total flavones of Abelmoschus manihot improve diabetic nephropathy by inhibiting the iRhom2/TACE signalling pathway activity in rats. Pharm Biol 56:1–11. https://doi.org/10.1080/13880209.2017.1412467

Lue S-H, Wellenius GA, Wilker EH, Mostofsky E, Mittleman MA (2013) Residential proximity to major roadways and renal function. J Epidemiol Community Health 67:629–634. https://doi.org/10.1136/jech-2012-202307

Lundstedt S, White PA, Lemieux CL, Lynes KD, Lambert IB, Öberg L, Haglund P, Tysklind M (2007) Sources, fate, and toxic hazards of oxygenated polycyclic aromatic hydrocarbons (PAHs) at PAH-contaminated sites. AMBIO J Hum Environ 36:475–486. https://doi.org/10.1579/0044-7447(2007)36[475:SFATHO]2.0.CO;2

Ma Z, Hu X, Sayer AM, Levy R, Zhang Q, Xue Y, Tong S, Bi J, Huang L, Liu Y (2015) Satellite-based spatiotemporal trends in PM2. 5 concentrations: China, 2004–2013. Environ Health Perspect 124:184–192. https://doi.org/10.1289/ehp.1409481

Martinelli N, Olivieri O, Girelli D (2013) Air particulate matter and cardiovascular disease: a narrative review. Eur J Intern Med 24:295–302. https://doi.org/10.1016/j.ejim.2013.04.001

Mehta AJ, Zanobetti A, Bind M-AC, Kloog I, Koutrakis P, Sparrow D, Vokonas PS, Schwartz JD (2016) Long-term exposure to ambient fine particulate matter and renal function in older men: the veterans administration normative aging study. Environ Health Perspect 124:1353–1360. https://doi.org/10.1289/ehp.1510269

Milojevic A, Wilkinson P, Armstrong B, Bhaskaran K, Smeeth L, Hajat S (2014) Short-term effects of air pollution on a range of cardiovascular events in England and Wales: case-crossover analysis of the MINAP database, hospital admissions and mortality. Heart 100:1093–1098. https://doi.org/10.1136/heartjnl-2013-304963

Monn C, Becker S (1999) Cytotoxicity and induction of proinflammatory cytokines from human monocytes exposed to fine (PM2. 5) and coarse particles (PM10–2.5) in outdoor and indoor air. Toxicol Appl Pharmacol 155:245–252. https://doi.org/10.1006/taap.1998.8591

Morakinyo OM, Mokgobu MI, Mukhola MS, Hunter RP (2016) Health outcomes of exposure to biological and chemical components of inhalable and respirable particulate matter. Int J Environ Res Public Health 13:592. https://doi.org/10.3390/ijerph13060592

Mukherjee A, Agrawal M (2017) A global perspective of fine particulate matter pollution and its health effects. In: Reviews of environmental contamination and toxicology. Springer, Berlin, pp 5–51. https://doi.org/10.1007/398_2017_3

National Clean Air Programme (2019) Extensive consultation with all the relevant stakeholders foundation for formulation of NCAP: Dr. Harsh Vardhan. https://pib.gov.in/PressReleasePage.aspx?PRID=1559384

Nemmar A, Al-Salam S, Zia S, Yasin J, Al Husseni I, Ali BH (2009) Diesel exhaust particles in the lung aggravate experimental acute renal failure. Toxicol Sci 113:267–277. https://doi.org/10.1093/toxsci/kfp222

Niazi ZR, Silva GC, Ribeiro TP, León-González AJ, Kassem M, Mirajkar A, Alvi A, Abbas M, Zgheel F, Schini-Kerth VB (2017) EPA: DHA 6: 1 prevents angiotensin II-induced hypertension and endothelial dysfunction in rats: role of NADPH oxidase-and COX-derived oxidative stress. Hypertens Res 40:966. https://doi.org/10.1038/hr.2017.72

Papademetriou V, Zaheer M, Doumas M, Lovato L, Applegate WB, Tsioufis C, Mottle A, Punthakee Z, Cushman WC, AS Group (2016) Cardiovascular outcomes in action to control cardiovascular risk in diabetes: Impact of blood pressure level and presence of kidney disease. Am J Nephrol 43:271–280. https://doi.org/10.1159/000446122

Parikh NI, Hwang S-J, Larson MG, Meigs JB, Levy D, Fox CS (2006) Cardiovascular disease risk factors in chronic kidney disease: overall burden and rates of treatment and control. Arch Intern Med 166:1884–1891. https://doi.org/10.1001/archinte.166.17.1884

Pérez N, Pey J, Querol X, Alastuey A, López J, Viana M (2008) Partitioning of major and trace components in PM10–PM2. 5–PM1 at an urban site in Southern Europe. Atmos Environ 42:1677–1691. https://doi.org/10.1016/j.atmosenv.2007.11.034

Pérez N, Pey J, Cusack M, Reche C, Querol X, Alastuey A, Viana M (2010) Variability of particle number, black carbon, and PM10, PM2. 5, and PM1 levels and speciation: influence of road traffic emissions on urban air quality. Aerosol Sci Tech 44:487–499. https://doi.org/10.1080/02786821003758286

Pfau JC, Brown JM, Holian A (2004) Silica-exposed mice generate autoantibodies to apoptotic cells. Toxicology 195:167–176. https://doi.org/10.1016/j.tox.2003.09.011

Pfau JC, Sentissi JJ, Sa L, Calderon-Garcidueñas L, Brown JM, Blake DJ (2008) Asbestos-induced autoimmunity in C57BL/6 mice. J Immunotoxicol 5:129–137. https://doi.org/10.1080/15476910802085756

Pollution Issues (2006) Indoor Air Pollution. http://www.pollutionissues.com/Ho-Li/Indoor-Air-Pollution.html

Power AL, Tennant RK, Jones RT, Tang Y, Du J, Worsley AT, Love J (2018) Monitoring impacts of urbanisation and industrialisation on air quality in the Anthropocene using urban pond sediments. Frontiers in Earth Science 6:131. https://doi.org/10.3389feart.2018.00131

Renshaw N, Dissard R, Beagley J (2019) NCD alliance annual report 2018. https://ncdalliance.org/sites/default/files/resource_files/NCDA_AnnualReport%202018_FV_LowRes.pdf

Rui W, Guan L, Zhang F, Zhang W, Ding W (2016) PM2. 5-induced oxidative stress increases adhesion molecules expression in human endothelial cells through the ERK/AKT/NF-κB-dependent pathway. J Appl Toxicol 36:48–59. https://doi.org/10.1002/jat.3143

Samake A, Uzu G, Martins J, Calas A, Vince E, Parat S, Jaffrezo J (2017) The unexpected role of bioaerosols in the oxidative potential of PM. Sci Rep 7:10978. https://doi.org/10.1038/s41598-017-11178-0

Santana PT, Martel J, Lai H-C, Perfettini J-L, Kanellopoulos JM, Young JD, Coutinho-Silva R, Ojcius DM (2016) Is the inflammasome relevant for epithelial cell function? Microbes Infect 18:93–101. https://doi.org/10.1016/j.micinf.2015.10.007

Sharma H, Jain V, Khan ZH (2007) Characterization and source identification of polycyclic aromatic hydrocarbons (PAHs) in the urban environment of Delhi. Chemosphere 66:302–310. https://doi.org/10.1016/j.chemosphere.2006.05.003

Singh A, Kamal R, Mudiam MKR, Gupta MK, Satyanarayana GNV, Bihari V, Shukla N, Khan AH, Kesavachandran CN (2016) Heat and PAHs emissions in indoor kitchen air and its impact on kidney dysfunctions among kitchen workers in Lucknow, North India. PLoS One 11: e0148641. https://doi.org/10.1371/journal.pone.0148641

Sørensen M, Schins RP, Hertel O, Loft S (2005) Transition metals in personal samples of PM2. 5 and oxidative stress in human volunteers. Cancer Epidemiol Prev Biomark 14:1340–1343. https://doi.org/10.1158/1055-9965

Srikanth P, Sudharsanam S, Steinberg R (2008) Bio-aerosols in indoor environment: composition, health effects and analysis. Indian J Med Microbiol 26:302. https://doi.org/10.4103/0255-0857.43555

Stojan G, Curriero F, Kvit A, Petri MA (2019) Environmental and atmospheric factors in systemic lupus erythematosus: a regression analysis. Ann Rheum Dis 78(Suppl 2). https://doi.org/10.1136/annrheumdis-2019-eular.5962

Sumida K, Molnar MZ, Potukuchi PK, George K, Thomas F, Lu JL, Yamagata K, Kalantar-Zadeh K, Kovesdy CP (2017) Changes in albuminuria and subsequent risk of incident kidney disease. Clin J Am Soc Nephrol 12:1941–1949. https://doi.org/10.2215/CJN.02720317

Tao J, Wang Y, Wei J, Zhang R, Li T (2016) Characteristics of organic and elemental carbon in the indoor/outdoor air PM_ (2. 5) in Beijing. Wei sheng yan jiu J Hyg Res 45:795–806

The U.S. Environmental Protection Agency (2012) Revised air quality standards for particle pollution and updates to the air quality index (AQI). https://www.epa.gov/sites/production/files/2016-04/documents/2012_aqi_factsheet.pdf

Thomaidis NS, Bakeas EB, Siskos PA (2003) Characterization of lead, cadmium, arsenic and nickel in PM2. 5 particles in the Athens atmosphere, Greece. Chemosphere 52:959–966. https://doi.org/10.1016/S0045-6535(03)00295-9

Tonelli M, Muntner P, Lloyd A, Manns BJ, Klarenbach S, Pannu N, James MT, Hemmelgarn BR, AKD Network (2012) Risk of coronary events in people with chronic kidney disease compared with those with diabetes: a population-level cohort study. Lancet 380:807–814. https://doi.org/10.1016/S0140-6736(12)60572-8

Vouk VB, Piver WT (1983) Metallic elements in fossil fuel combustion products: amounts and form of emi Zhang, Y., Ji, X., Ku, T., Li, G., & Sang, N. (2016). Heavy metals bound to fine particulate matter from northern China induce season-dependent health risks: a study based on myocardial toxicity. Environ Pollut 216:380–390. https://doi.org/10.1016/j.envpol.2016.05.072

Wang J, Li X, Jiang N, Zhang W, Zhang R, Tang X (2015) Long term observations of PM2. 5-associated PAHs: Comparisons between normal and episode days. Atmos Environ 104:228–236. https://doi.org/10.1016/j.atmosenv.2015.01.026

Wang T, Yuan Y, Zou H, Yang J, Zhao S, Ma Y, Wang Y, Bian J, Liu X, Gu J (2016) The ER stress regulator Bip mediates cadmium-induced autophagy and neuronal senescence. Sci Rep 6:38091. https://doi.org/10.1038/srep38091

Wang Q, Jiang N, Yin S, Li X, Yu F, Guo Y, Zhang R (2017a) Carbonaceous species in PM2. 5 and PM10 in urban area of Zhengzhou in China: seasonal variations and source apportionment. Atmos Res 191:1–11. https://doi.org/10.1016/j.atmosres.2017.02.003

Wang W, Deng Z, Feng Y, Liao F, Zhou F, Feng S, Wang X (2017b) PM2. 5 induced apoptosis in endothelial cell through the activation of the p53-bax-caspase pathway. Chemosphere 177:135–143. https://doi.org/10.1016/j.chemosphere.2017.02.144

Weichenthal SA, Lavigne E, Evans GJ, Godri Pollitt KJ, Burnett RT (2016) Fine particulate matter and emergency room visits for respiratory illness. Effect modification by oxidative potential. Am J Respiratory Crit Care Med 194:577–586. https://doi.org/10.1164/rccm.201512-2434OC

World Development Indicators (2017) EN.ATM.PM25.MC.T1.ZS, SP.POP.TOTL. https://databank.worldbank.org/WHO-Interim-Target-1/id/8030eb3e

World Health Organization (2006) WHO Air quality guidelines for particulate matter, ozone, nitrogen dioxide and sulfur dioxide. https://apps.who.int/iris/bitstream/handle/10665/69477/WHO_SDE_PHE_OEH_06.02_eng.pdf;jsessionid=07757F4EC9D55DE3904B89AFE1D9D155?sequence=1

World Health Organization (2014) WHO global status report on NCDs, 2014. http://apps.who.int/iris/bitstream/10665/148114/1/9789241564854_eng.pdf

World Health Organization (2018a) 9 out of 10 people worldwide breathe polluted air, but more countries are taking action. https://www.who.int/zh/news-room/detail/02-05-2018-9-out-of-10-people-worldwide-breathe-polluted-air-but-more-countries-are-taking-action

World Health Organization (2018b) Ambient (outdoor) air quality and health. https://www.who.int/news-room/fact-sheets/detail/ambient-(outdoor)-air-quality-and-health

World Health Organization (2018c) WHO Global Ambient Air Quality Database (update 2018). https://www.who.int/airpollution/data/cities/en/

Wu MY, Lo WC, Chao CT, Wu MS, Chiang CK (2020) Association between air pollutants and development of chronic kidney disease: a systematic review and meta-analysis. Sci Total Environ 706:135522. https://doi.org/10.1016/j.envpol.2020.114154

Xu X, Wang G, Chen N, Lu T, Nie S, Xu G, Zhang P, Luo Y, Wang Y, Wang X (2016) Long-term exposure to air pollution and increased risk of membranous nephropathy in China. J Am Soc Nephrol 27:3739–3746. https://doi.org/10.1681/ASN.2016010093

Xu X, Wang H, Xing C, Gu Y, Liu S, Xu H, Hu M, Song L (2017) IRE1α/XBP1s branch of UPR links HIF1α activation to mediate ANGII-dependent endothelial dysfunction under particulate matter (PM) 2.5 exposure. Sci Rep 7:13507. https://doi.org/10.1038/s41598-017-13156-y

Xu M-X, Qin Y-T, Ge C-X, Gu T-T, Lou D-S, Li Q, Hu L-F, Li Y-Y, Yang W-W, Tan J (2018) Activated iRhom2 drives prolonged PM2. 5 exposure-triggered renal injury in Nrf2-defective mice. Nanotoxicology 12:1045–1067. https://doi.org/10.1080/17435390.2018.1513093

Yan Y-H, Chou CC-K, Wang J-S, Tung C-L, Li Y-R, Lo K, Cheng T-J (2014) Subchronic effects of inhaled ambient particulate matter on glucose homeostasis and target organ damage in a type 1 diabetic rat model. Toxicol Appl Pharmacol 281:211–220. https://doi.org/10.1016/j.taap.2014.10.005

Yang YR, Chen YM, Chen SY, Chan CC (2017) Associations between long-term particulate matter exposure and adult renal function in the Taipei metropolis. Environ Health Perspect 125 (4):602–607. https://doi.org/10.1289/EHP302

Yang S, Li R, Shu X, Li Q (2018) Environmental pollutants and diabetic kidney disease (DKD). Med Res Arch 6. https://doi.org/10.18103/mra.v6i4.1770

Yunker MB, Macdonald RW, Vingarzan R, Mitchell RH, Goyette D, Sylvestre S (2002) PAHs in the Fraser River basin: a critical appraisal of PAH ratios as indicators of PAH source and composition. Org Geochem 33:489–515. https://doi.org/10.1016/S0146-6380(02)00002-5

Zanobetti A, Franklin M, Koutrakis P, Schwartz J (2009) Fine particulate air pollution and its components in association with cause-specific emergency admissions. Environ Health 8:58. https://doi.org/10.1186/1476-069X-8-58

Zhai S, Jacob DJ, Wang X, Shen L, Li K, Zhang Y, Gui K, Zhao T, Liao H (2019) Fine particulate matter (PM2.5) trends in China, 2013–2018: separating contributions from anthropogenic emissions and meteorology. Atmos Chem Phys 19:11031–11041. https://doi.org/10.5194/acp-19-11031-2019

Zhang Y-L, Cao F (2015) Fine particulate matter (PM 2.5) in China at a city level. Sci Rep 5:14884. https://doi.org/10.1038/srep14884

Zhang L, Wang F, Wang L, Wang W, Liu B, Liu J, Chen M, He Q, Liao Y, Yu X (2012) Prevalence of chronic kidney disease in China: a cross-sectional survey. Lancet 379:815–822. https://doi.org/10.1016/S0140-6736(12)60033-6

Zhang Y, Ji X, Ku T, Li G, Sang N (2016) Heavy metals bound to fine particulate matter from northern China induce season-dependent health risks: a study based on myocardial toxicity. Environ Pollut 216:380–390. https://doi.org/10.1016/j.envpol.2016.05.072

Zhang L, Xu T, Pi Z, Zheng M, Song F, Liu Z (2017) Effects of the ambient fine particulate matter (PM2. 5) exposure on urinary metabolic profiles in rats using UPLC-Q-TOF-MS. Chin Chem Lett. https://doi.org/10.1016/j.cclet.2017.11.019

Zhang Y, Li Q, Fang M, Ma Y, Liu N, Yan X, Zhou J, Li F (2018) The kidney injury induced by short-term PM2. 5 exposure and the prophylactic treatment of essential oils in BALB/c Mice. Oxidat Med Cell Longevity 2018. https://doi.org/10.1155/2018/9098627

Zhang Y, Liu D, Liu Z (2020) The benzo [b] fluoranthene in the atmospheric fine particulate matter induces mouse glomerular podocytes injury via inhibition of autophagy. Ecotoxicol Environ Saf 195:110403. https://doi.org/10.1016/j.ecoenv.2020.110403

Zheng L, Liu S, Zhuang G, Xu J, Liu Q, Zhang X, Deng C, Guo Z, Zhao W, Liu T (2017) Signal transductions of BEAS-2B cells in response to carcinogenic PM2. 5 exposure based on a microfluidic system. Anal Chem 89:5413–5421. https://doi.org/10.1021/acs.analchem.7b00218

Zhong J, Karlsson O, Wang G, Li J, Guo Y, Lin X, Zemplenyi M, Sanchez-Guerra M, Trevisi L, Urch B (2017) B vitamins attenuate the epigenetic effects of ambient fine particles in a pilot human intervention trial. Proc Natl Acad Sci 114:3503–3508. https://doi.org/10.1073/pnas.1618545114

Zhu X, Liu Y, Chen Y, Yao C, Che Z, Cao J (2015) Maternal exposure to fine particulate matter (PM 2.5) and pregnancy outcomes: a meta-analysis. Environ Sci Pollut Res 22:3383–3396. https://doi.org/10.1007/s11356-014-3458-7

Correction to: Occurrence of Levonorgestrel in Water Systems and Its Effects on Aquatic Organisms: A Review

Ana Lourdes Oropesa and Laura Guimarães

Correction to:
Chapter "Occurrence of Levonorgestrel in Water Systems and Its Effects on Aquatic Organisms: A Review" in: Ana Lourdes Oropesa and Laura Guimarães, *Reviews of Environmental Contamination and Toxicology,* https://doi.org/10.1007/398_2020_44

The author has requested to include the below Acknowledgement section at the end of the chapter.

Acknowledgements This work was supported by UIDB/04423/2020 and UIDP/04423/2020, through FCT (Portuguese Foundation for the Science and Technology), and by the REWATER project (FCT/UEFISCDI/FORMAS, WaterJPI/0008/2016) financed under the ERA-NET Cofund WaterWorks2015 (Water JPI)."

The original article was corrected.

The updated online version of this chapter can be found at
https://doi.org/10.1007/398_2020_44

A. L. Oropesa (✉)
Toxicology Area, University of Extremadura, Caceres, Spain

BRILiC – Biotechnology Research Institute in Livestock and Cynegetic, University of Extremadura, Caceres, Spain
e-mail: aoropesa@unex.es

P. de Voogt (ed.), *Reviews of Environmental Contamination and Toxicology Volume 254*,
Reviews of Environmental Contamination and Toxicology 254,
https://doi.org/10.1007/398_2020_52

GPSR Compliance

The European Union's (EU) General Product Safety Regulation (GPSR) is a set of rules that requires consumer products to be safe and our obligations to ensure this.

If you have any concerns about our products, you can contact us on ProductSafety@springernature.com

In case Publisher is established outside the EU, the EU authorized representative is:

Springer Nature Customer Service Center GmbH
Europaplatz 3
69115 Heidelberg, Germany

Batch number: 10777422

Printed by Printforce, the Netherlands